26

Heinz Georg Schuster

Deterministic Chaos

© VCH Verlagsgesellschaft mbH, D-6940 Weinheim (Federal Republic of Germany), 1988

Distribution

VCH Verlagsgesellschaft, P.O. Box 1260/1280, D-6940 Weinheim (Federal Republic of Germany)

Switzerland: VCH Verlags-AG, P.O. Box, CH-4020 Basel (Switzerland)

Great Britain and Ireland: VCH Publishers (UK) Ltd., 8 Wellington Court, Wellington Street, Cambridge CB1 1HW (Great Britain)

USA and Canada: VCH Publishers, Suite 909, 220 East 23rd Street, New York, NY 10010-4606 (USA)

ISBN 3-527-26862-6 (VCH Verlagsgesellschaft)   ISBN 0-89573-611-X (VCH Publishers)

Heinz Georg Schuster

Deterministic Chaos

An Introduction

Second revised edition

Prof. Dr. Heinz Georg Schuster
Institut für Theoretische Physik
und Sternwarte der Universität
Olshausenstraße — Physikzentrum
D-2300 Kiel
Federal Republic of Germany

Editorial Director: Walter Greulich
Production Manager: Heidi Lenz

Deutsche Bibliothek, Cataloguing-in-Publication Data

Schuster, Heinz G.:
Deterministic chaos: an introd. / Heinz Georg Schuster. — 2., rev. ed. — Weinheim; Basel; Cambridge; New York, NY: VCH, 1988.
   ISBN 3-527-26862-6 (Weinheim ...)
   ISBN 0-89573-611-X (Cambridge ...)

British Library Cataloguing in Publication Data

A CIP catalogue record for this book is available from the British Library

Library of Congress Card No. 87-7189

© VCH Verlagsgesellschaft mbH, D-6940 Weinheim (Federal Republic of Germany), 1988

Composition: Filmsatz Unger, D-6940 Weinheim
Printing: betz-druck gmbh, D-6100 Darmstadt 12
Bookbinding: Georg Kränkl, D-6148 Heppenheim
Printed in the Federal Republic of Germany

Dedicated to Gaby

Preface

Daily experience shows that for many physical systems small changes in the initial conditions lead to small changes in the outcome. If we drive a car and turn the steering wheel only a little, our course will differ only slightly from that which the car would have taken without this change.

But there are cases for which the opposite of this rule is true: For a coin which is placed on its rim, a slight touch is sufficient to determine the side on which it will fall. Thus the sequence of heads and tails which we obtain when tossing a coin exhibits an irregular or chaotic behavior in time, because extremely small changes in the initial conditions can lead to completely different outcomes.

It has become clear in recent years, partly triggered by the studies of nonlinear systems using high-speed computers, that a *sensitive dependence on the initial conditions*, which results in a chaotic time-behavior, is by no means exceptional but a *typical property of many systems*. Such behavior has, for example, been found in periodically stimulated cardiac cells, in electronic circuits, at the onset of turbulence in fluids and gases, in chemical reactions, in lasers, etc. Mathematically, all nonlinear dynamical systems with more than two degrees of freedom, i.e. especially many biological, meteorological or economic models, can display chaos and, therefore, *become unpredictable over longer time scales*.

"Deterministic chaos" is now a very active field of research with many exciting results. Methods have been developed to classify different types of chaos, and it has been discovered that many systems show, as a function of an external control parameter, similar transitions from order to chaos. This universal behavior is reminiscent of ordinary second-order phase transitions, and the introduction of renormalization and scaling methods from statistical mechanics has brought new perspectives into the study of deterministic chaos.

It is the aim of this book to provide a self-contained introduction to this field from a physicist's point of view. The book grew out of a series of lectures, which I gave during the summer terms of 1982 and 1983 at the University of Frankfurt, and it requires no knowledge which a graduate student in physics would not have. A glance on the table of contents shows that new concepts such as the Kolmogorov entropy, strange attractors, etc., or new techniques such as the functional renormalization group are introduced on an elementary level. On the other hand, I hope that there is enough material for research workers who want to know, for example, how deterministic chaos can be distinguished experimentally from white noise, or who want to learn how to apply their knowledge about equilibrium phase transitions to the study of (nonequilibrium) transitions from order to chaos.

During the preparation of this book the manuscripts, preprints and discussion remarks of G. Eilenberger, K. Kehr, H. Leschke, W. Selke, and M. Schmutz were of great help. P. Bergé, M. Dubois, W. Lauterborn, W. Martienssen, G. Pfister and their coworkers supplied several, partly unpublished pictures of their experiments. H. O. Peitgen, P. H. Richter and their group gave the permission to include some of their most fascinating computer pictures into this book (see cover and Sect. 5.4). All contributions are gratefully appreciated. Furthermore, I want to thank W. Greulich, D. Hackenbracht, M. Heise, L. L. Hirst, R. Liebmann, I. Neil, and especially I. Procaccia for carefully reading parts of the manuscript and for useful criticism and comments. I also acknowledge illuminating discussions with V. Emery, P. Grassberger, D. Grempel, S. Grossmann, S. Fishman, and H. Horner.

It is a pleasure to thank R. Hornreich for the kind hospitality extended to me during a stay at the Weizmann Institute, where several chapters of this book were written, with the support of the Minerva foundation.

Last but not least, I thank Mrs. Boffo and Mrs. Knolle for their excellent assistance in preparing the illustrations and the text.

Frankfurt, October 1984   H. G. Schuster

Preface to the Second Edition

This is a revised and updated version of the first edition, to which new sections on sensitive parameter dependence, fat fractals, characterization of attractors by scaling indices, the Farey tree, and the notion of global universality have been added. I thank P. C. T. de Boer, J. L. Grant, P. Grassberger, W. Greulich, F. Kaspar, K. Pawelzik, K. Schmidt, and S. Smid for helpful hints and remarks, and Mrs. Adlfinger and Mrs. Boffo for their patient help with the manuscript.

Kiel, August 1987   H. G. Schuster

Preface

Contents

Legends to Plates I–XVII

Many of these plates are part of chapter 5. Accordingly, references mentioned in the legends are to be found on pages 257–258.

I   *Biperiodic flow in a Bénard experiment:* Figs. 1–8 show interferometric pictures of a Bénard cell in the biperiodic régime; that is, there are two incommensurate frequencies in the power spectrum (see also pages 9–11). The time between successive pictures is 10 s. The first period lasts 40 s after which the "mouth" in the middle of the pictures repeats itself (see Figs. 1 and 5). But the details, e. g. in the upper right corners of Figs. 1 and 5 are not the same; that is, the motion is *not* simply periodic. (From a film taken by P. Bergé and M. Dubois, CEN Saclay, Gif-sur-Yvette, France.)

II   *Nonlinear electronic oscillator* (see also Fig. 46 on page 75): The current-versus-voltage phase portraits (at the nonlinear diode) are shown on the oscilloscope screen. For increasing driving voltage one observes the period-doubling route. The nonlinearity of the diode that has been used in this experiment differs from eq. (3.121). (Picture taken by W. Meyer-Ilse, after Klinker et al., 1984.)

III   *Taylor instability:* a) Formation of rolls, b) the rolls start oscillating, c) a more complicated oscillatory motion, d) chaos. (After Pfister, 1984; see also pages 152–154.)

IV   *Disturbed heartbeats:* The voltage difference (black) across the cell membrane of one cell of an aggregate of heart cells from embryonic chicken shows a) phase looking with the stimulating pulse and b) irregular dynamics, displaying escape or interpolation beats if the time between successive periodic stimuli (red) is changed from 240 ms in a) to 560 ms in b). (After Glass et al., 1983; see also page 177.)

V   *Chaotic electrical conduction in BSN crystals:* The birefringence pattern of a ferroelectric BSN crystal shows domain walls which mirror the charge transport near the onset of chaos (see also Fig. 115 on page 176). For clarity, the dark lines in the original pattern have been redrawn in red. (After Martin et al., 1984.)

1

2

3

4

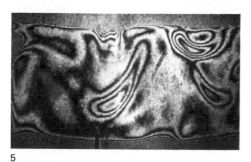

5

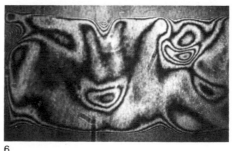

6

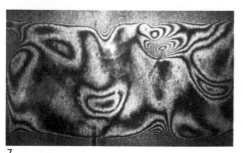

7

8

Plate I XV

II

III

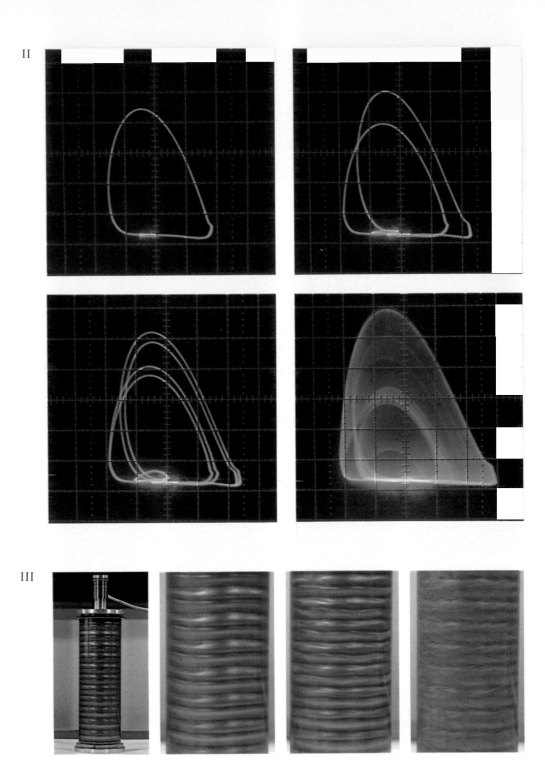

Plates II and III

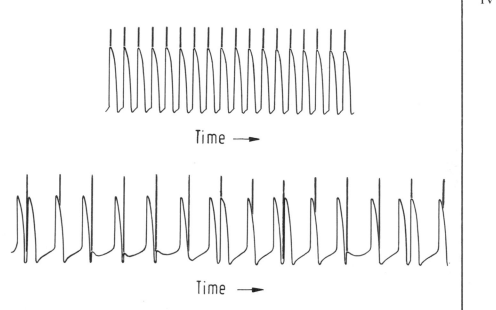

Time ⟶

Time ⟶

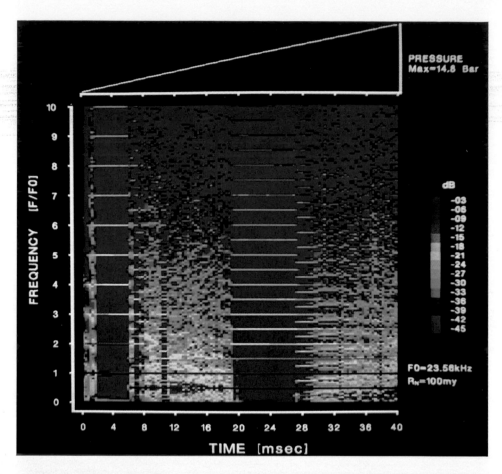

PRESSURE
Max=14.8 Bar

FREQUENCY [F/F0]

F0=23.56kHz
R$_N$=100my

TIME [msec]

dB
-03
-06
-09
-12
-15
-18
-21
-24
-27
-30
-33
-36
-39
-42
-45

VII

XVIII   *Plates VI and VII*

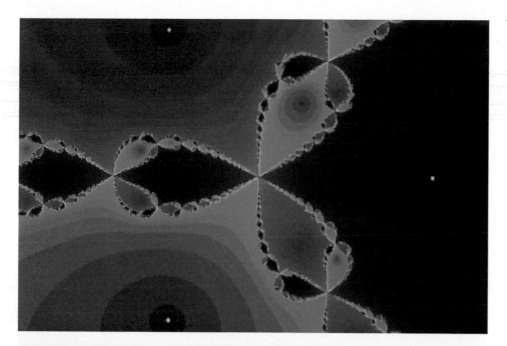

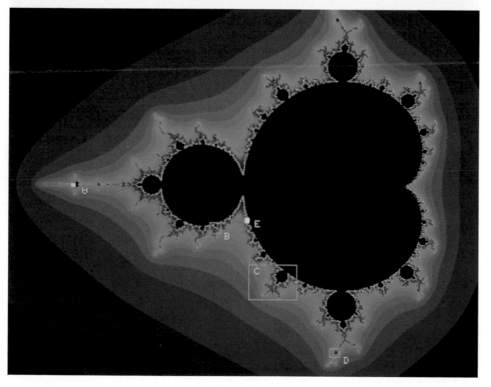

X

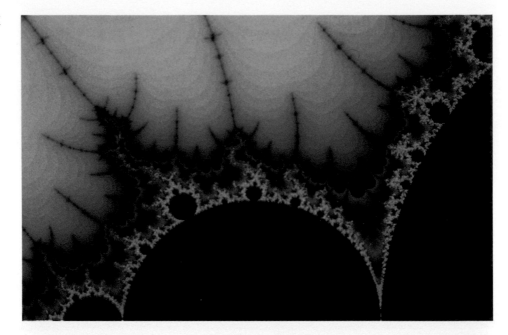

XI

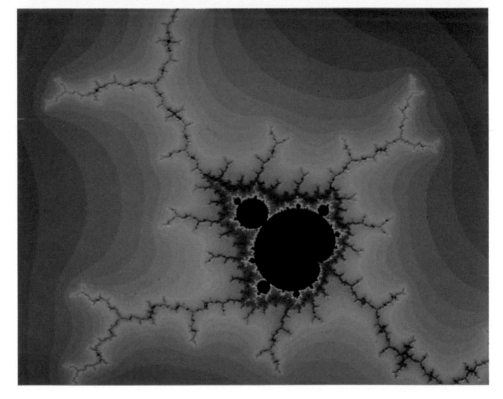

XX   *Plates X and XI*

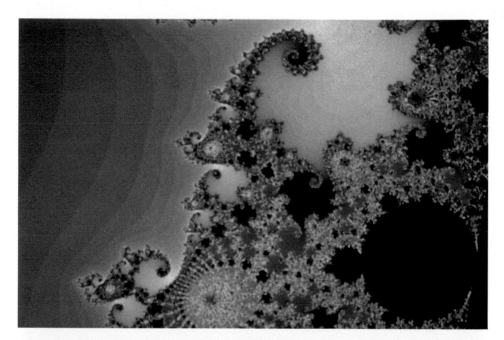

XIV

XV

XXII   *Plates XIV and XV*

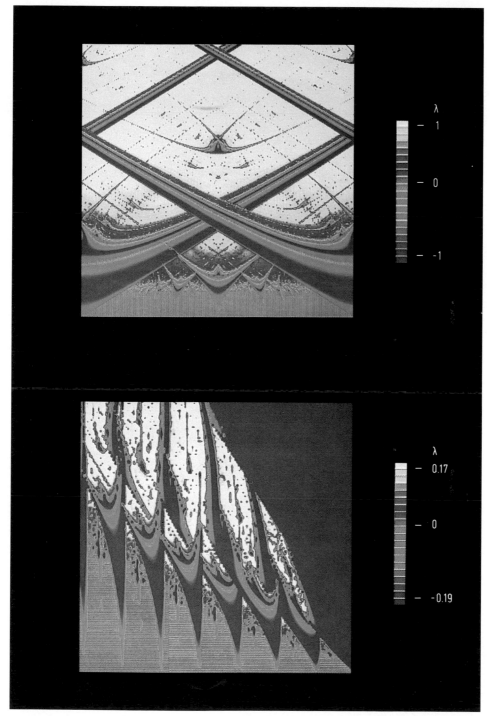

Ante mare et terras et, quod tegit omnia, caelum
Unus erat toto naturae vultus in orbe,
Quem dixere Chaos, rudis indigestaque moles
Nec quicquam nisi pondus iners congestaque
eodem
Non bene iunctarum discordia semina rerum.

Ovid

Introduction

It seems appropriate to begin a book which is entitled "Deterministic Chaos" with an explanation of both terms. According to the Encyclopaedia Britannica the word *"chaos"* is derived from the Greek "χαος" and originally meant the infinite empty space which existed before all things. The later Roman conception interpreted chaos as the original crude shapeless mass into which the Architect of the world introduces order and harmony. In modern usage which we will adopt here, chaos denotes a state of disorder and irregularity.

In the following, we shall consider physical systems whose time dependence is *deterministic,* i. e. there exists a prescription, either in terms of differential or difference equations, for calculating their future behavior from given initial conditions. One could assume naively that deterministic motion (which is, for example, generated by continuous differential equations) is rather regular and far from being chaotic because successive states evolve continuously from each other. But it was already discovered at the turn of the century by the mathematician H. Poincaré (1892) that certain mechanical systems whose time evolution is governed by Hamilton's equations could display chaotic motion. Unfortunately, this was considered by many physicists as a mere curiosity, and it took another 70 years until in 1963 the meteorologist E. N. Lorenz found that even a simple set of three coupled, first order, nonlinear differential equations can lead to completely chaotic trajectories. Lorenz's paper, the general importance of which is recognized today, was also not widely appreciated until many years after its publication. He discovered one of the first examples of *deterministic chaos* in dissipative systems.

In the following, deterministic chaos denotes the irregular or chaotic motion that is generated by nonlinear systems whose dynamical laws uniquely determine the time evolution of a state of the system from a knowledge of its previous history. In recent years − due to new theoretical results, the availability of high speed computers, and refined experimental techniques − it has become clear that this phenomenon is abundant in nature and has far-reaching consequences in many branches of science (see the long list in Table 1 which is far from complete).

We note that nonlinearity is a necessary, but not a sufficient condition for the generation of chaotic motion. (Linear differential or difference equations can be solved by Fourier transformation and do not lead to chaos.)

Table 1: Some nonlinear systems which display deterministic chaos. (For numerals, see "References" on page 247.)

Forced pendulum [1]
Fluids near the onset of turbulence [2]
Lasers [3]
Nonlinear optical devices [4]
Josephson junctions [5]
Chemical reactions [6]
Classical many-body systems (three-body problem) [7]
Particle accelerators [8]
Plasmas with interacting nonlinear waves [9]
Biological models for population dynamics [10]
Stimulated heart cells (see Plate IV at the beginning of the book) [11]

The observed chaotic behavior in time is neither due to external sources of noise (there are none in the Lorenz equations) nor to an infinite number of degrees of freedom (in Lorenz's system there are only three degrees of freedom) nor to the uncertainty associated with quantum mechanics (the systems considered are purely classical). The actual source of irregularity is the property of the nonlinear system of separating initially close trajectories exponentially fast in a bounded region of phase space (which is, e. g., three-dimensional for Lorenz's system).

It becomes therefore practically impossible to predict the long-time behavior of these systems, because in practice one can only fix their initial conditions with finite accuracy, and errors increase exponentially fast. If one tries to solve such a nonlinear system on a computer, the result depends for longer and longer times on more and more digits in the (irrational) numbers which represent the initial conditions. Since the digits in irrational numbers (the rational numbers are of measure zero along the real axis) are irregularly distributed, the trajectory becomes chaotic.

Lorenz called this *sensitive dependence on the initial conditions* the butterfly effect, because the outcome of his equations (which describe also, in a crude sense, the flow of air in the earth's atmosphere, i. e. the problem of the weather forecasting) could be changed by a butterfly flapping his wings. This also seems to be confirmed sometimes by daily experience.

The results described above immediately raise a number of fundamental questions:

Can one predict (e. g. from the form of the corresponding differential equations) whether or not a given system will display deterministic chaos?

Can one specify the notion of chaotic motion more mathematically and develop quantitative measures for it?

What is the impact of these findings on different branches in physics?

Does the existence of deterministic chaos imply the end of long-time predictability in physics for some nonlinear systems, or can one still learn something from a chaotic signal?

The last question really goes to the fundaments of physics, namely the problem of predictability. The shock which was associated with the discovery of deterministic chaos has therefore been compared to that which spread when it was found that quantum mechanics only allows statistical predictions.

Those questions mentioned above, to which some answers already exist, will be discussed in the remainder of this book. It should be clear, however, that there are still many more unsolved than solved problems in this relatively new field.

The rest of the introduction takes the form of a short survey which summarizes the contents of this book.

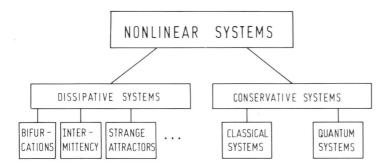

Fig. 1: Classification of systems which display deterministic chaos. (We consider in the following only classical dissipative systems, i. e. no quantum systems with dissipation.)

Fig. 1 shows that one has to distinguish between deterministic chaos in dissipative systems (e.g. a forced pendulum with friction) and conservative systems (e.g. planetary motion which is governed by Hamilton's equations).

The first six chapters are devoted to dissipative systems. We begin with a review of some representative experiments in which deterministic chaos has been observed by different methods. As a next step, we explain the mechanism which leads to deterministic chaos for a simple model system and develop quantitative measures to characterize a chaotic signal. This allows us to distinguish different types of chaos, and we then show that, up to now, there are at least three routes or transitions in which nonlinear systems can become chaotic if an external control parameter is varied. Interestingly enough, all these routes can be realized experimentally, and they show a fascinating universal behavior which is reminiscent of the universality found in second-order equilibrium phase transitions. (Note that the transitions to chaos in dissipative systems only occur when the system is driven externally, i. e. is open.) In this context, universality means that there are basic properties of the system (such as critical exponents near the transition to chaos) that depend only on some global features of the system (for example, the dimensionality).

The most recent route to chaos has been found by Grossmann and Thomae (1977), Feigenbaum (1978), and Coullett and Tresser (1978). They considered a simple difference equation which, for example, has been used to describe the time dependence of populations in biology, and found that the population oscillated in time between stable values (fixed points) whose number doubles at distinct values of an external parameter. This continues until the number of fixed points becomes infinite at a finite parameter value, where the variation in time of the population becomes irregular. Feigenbaum has shown, and this was a major achievement, that these results are not restricted to this special model but are in fact *universal* and hold for a large variety of physical, chemical, and biological systems. This discovery has triggered an explosion of theoretical and experimental activity in this field. We will study this route in chapter three and show that its universal properties can be calculated using the functional renormalization group method.

A second approach to chaos, the so-called intermittency route, has been discovered by Manneville and Pomeau (1979). Intermittency means that a signal which behaves regularly (or laminarly) in time becomes interrupted by statistically distributed periods of irregular motion (intermittent bursts). The average number of these bursts increases with the variation of an external control parameter until the motion becomes completely chaotic. It will be shown in chapter four that this route also has universal features and provides a universal mechanism for $1/f$-noise in nonlinear systems.

Yet a third possibility was found by Ruelle and Takens (1971) and Newhouse (1978). In the seventies they suggested a transition to turbulent motion which was different from that proposed much earlier by Landau (1944, 1959). Landau considered turbulence in time as the limit of an infinite sequence of instabilities (Hopf bifurcations) each of which creates a new basic frequency. However, Ruelle, Takens, and Newhouse showed that after only two instabilities in the third step the trajectory becomes attracted to a bounded region of phase space in which initially close trajectories separate exponentially such that the motion becomes chaotic. These particular regions of phase space are called *strange attractors*. We will explain this concept in Chapter 5, where we will also discuss several methods of extracting information about the structure of the attractor from the measured chaotic time signal. The Ruelle-Takens-Newhouse route is (as are the previous two routes) well verified experimentally, and we will present some experimental data which show explicitly the appearance of strange attractors in Chapter 6.

To avoid the confusion which might arise by the use of the word turbulence, we note that what is meant here, is only turbulence in *time*. The results of Ruelle, Takens, and Newhouse also concern the *onset* of turbulence or chaotic motion in time. It is in fact one of the *aims* (but not yet the result) of the study of deterministic chaos in hydrodynamic systems to understand the mechanisms for *fully developed* turbulence, which implies irregular behavior in *time and space*.

We now come to the second branch in Fig. 1, which denotes chaotic motion in conservative systems.

Many textbooks give the incorrect impression that most systems in classical mechanics can be integrated. But as mentioned above, Poincaré (1892) was already aware that, e. g., the nonintegrable three-body problem of classical mechanics can lead to completely chaotic trajectories. About sixty years later, Kolmogorov (1954), Arnold (1963), and Moser (1967) proved, in what is now known as the KAM theorem, that the motion in the phase space of classical mechanics is neither completely regular nor completely irregular, but that the type of trajectory depends sensitively on the chosen initial conditions. Thus, stable regular classical motion is the exception, contrary to what is implied in many texts.

A study of the long-time behavior of conservative systems, which will be discussed in Chapter 7, is of some interest because it touches on such questions as: Is the solar system stable? How can one avoid irregular motion in particle accelerators? Is the self-generated deterministic chaos of some Hamiltonian systems strong enough to prove the ergodic hypothesis? (The ergodic hypothesis lies at the foundations of classical statistical mechanics and implies that the trajectory uniformly covers the energetically allowed region of classical phase space such that time averages can be replaced by the average over the corresponding phase space.)

Finally, in the last chapter we consider the behavior of quantum systems whose classical limit displays chaos. Such investigations are important, for example, for the problem of photodissociation, where a molecule is kicked by laser photons, and one wants to know how the incoming energy spreads over the quantum levels. (The corresponding classical system could show chaos because the molecular forces are highly nonlinear.) For several examples we show that the finite value of Planck's constant leads, together with the boundary conditions, to an almost-periodic behavior of the quantum system even if the corresponding classical system displays chaos. Although the difference between integrable and nonintegrable (chaotic) classical systems is still mirrored in some properties of their quantum counterparts (for example in the energy spectra), many problems in this field remain unsolved.

1 Experiments and Simple Models

In the first part of this chapter, we review some experiments in which deterministic chaos has been detected by different methods. In the second part, we present some simple systems which exhibit chaos and which can be treated analytically.

1.1 Experimental Detection of Deterministic Chaos

In the following section, we will discuss the appearance of chaos in four representative systems.

Driven Pendulum

Let us first consider the surprisingly simple example of a periodically driven pendulum. Its equation of motion is

$$\ddot{\theta} + \gamma \dot{\theta} + \sin \theta = A \cos (\omega t) \tag{1.1}$$

where the dot denotes the derivative with respect to time t, γ is the damping constant, and the right hand side describes a driving torque with amplitude A and frequency ω. (The coefficients of $\ddot{\theta}$ and $\sin \theta$ have been normalized to unity by choosing appropriate units for t and A). This equation has been numerically integrated for different sets of parameters (A, ω, γ), and Fig. 2 shows that the variation of the angle θ with time simply "looks chaotic" if the amplitude A of the driving torque reaches a certain value A_c. This is a possible, but rather imprecise criterion for chaos.

Before we proceed to improvements, three comments are in order. First, we would like to recall the well-known fact that the linearized version of the pendulum equation can be integrated exactly and does not lead to chaos. The emergence of chaos in the

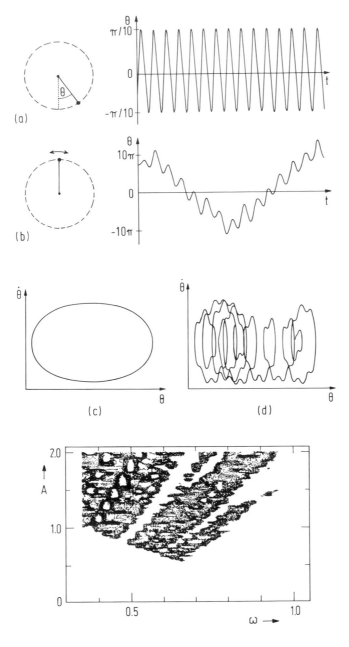

Fig. 2: Transition to chaos in a driven pendulum. a) Regular motion at small values of the amplitude A of the driving torque. b) Chaotic motion at $A = A_c$ (note the different scales for θ). c) and d) Regular and irregular trajectories in phase space $(\dot{\theta}, \theta)$ which correspond to a) and b). e) Phase diagram of the driven pendulum ($\gamma = 0.2$, $\theta(0) = 0$, $\dot{\theta}(0) = 0$). Black points denote parameter values (A, ω) for which the motion is chaotic. (After Bauer, priv. comm.)

solutions of eq. (1.1) is, therefore, due to the nonlinear term $\sin \theta$. Second, it follows from Fig. 2 b and 2 d that chaos sets in if the pendulum is driven over the summit where the system displays sensitive dependence on initial conditions (a tiny touch determines, at $\theta = \pi$, whether the pendulum makes a left or right turn). Third, we would like to point out that as a function of the parameters A and ω, the behavior of the pendulum switches rather wildly between regular and chaotic motion, as shown in Fig. 2 e.

Rayleigh-Bénard System in a Box

Chaotic motion means that the signal displays an irregular and aperiodic behavior in time. To distinguish between multiply periodic behavior (which can also look rather complicated) and chaos, it is often convenient to Fourier-transform the signal $x(t)$:

$$x(\omega) = \lim_{T \to \infty} \int_0^T dt\, e^{i\omega t} x(t) .\tag{1.2}$$

For multiply periodic motion, the power spectrum

$$P(\omega) \equiv |x(\omega)|^2 \tag{1.3}$$

consists only of discrete lines of the corresponding frequencies, whereas chaotic motion (which is completely aperiodic) is indicated by broad noise in $P(\omega)$ that is mostly located at low frequencies. Such a transition from periodic motion to chaos is presented in the second line of Table 2 which shows the power spectrum of the velocity of the liquid in the x-direction for a Bénard experiment.

In the Bénard experiment, a fluid layer (with a positive coefficient of volume expansion) is heated from below in a gravitational field, as shown in Fig. 3. The heated fluid at the bottom "wants" to rise, and the cold liquid at the top "wants" to fall, but these motions are opposed by viscous forces. For small temperature differences ΔT, viscosity wins; the liquid remains at rest and heat is transported by uniform heat conduction. This state becomes unstable at a critical value R_a of the Rayleigh number R (which is proportional to ΔT, see Appendix A), and a state of stationary convection rolls develops. If R increases, a transition to chaotic motion is observed beyond a second threshold R_c.

In order to avoid the appearance of complex spatial structures, actual experiments to detect chaos (in time) in a Rayleigh-Bénard system are usually performed in a small cell (see Fig. 3 c). The boundary conditions limit the number of rolls, that is the number of degrees of freedom that are counted by the number of Fourier components needed to describe the spatial structure of the fluid pattern. Besides ΔT, the observed dynamical behavior depends sensitively on the liquid chosen and on the linear dimensions (a, b, c) of the box (see, for example, Libchaber and Maurer, 1982).

Table 2 shows the power spectrum of the velocity in the x-direction, measured via the Doppler effect in light scattering experiments (Swinney and Gollub, 1978; see also

Table 2: Detection of chaos in simple systems.

System	Equation of Motion	Indication
Pendulum	$\ddot{\theta} + \gamma\dot{\theta} + g\sin\theta = A\cos\omega t$ $x = \theta,\ y = \dot{\theta},\ z = \omega t$ $\dot{x} = y$ $\dot{y} = -\gamma y - g\sin x + A\cos z$ $\dot{z} = \omega$	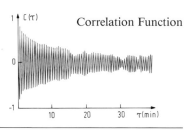 Signal
Bénard Experiment	$\dot{x} = -\sigma x + \sigma y$ $\dot{y} = rx - y - xz$ $\dot{z} = xy - bz$	Power Spectrum
Belousov-Zhabotinsky Reaction $Ce_2(SO_4)_3$ $\vdots$ Ce^{4+}	$\dot{\vec{x}} = \vec{F}(\vec{x}, \lambda)$ $\vec{x} = [c_1, c_2, \ldots c_d]$	Correlation Function
Hénon-Heiles System	$H = \dfrac{1}{2}\sum_{i=1}^{2}(p_i^2 + q_i^2) +$ $+ q_1^2 q_2 - \dfrac{1}{3}q_2^3$ $\dot{\vec{p}} = -\dfrac{\partial H}{\partial \vec{q}},\ \dot{\vec{q}} = \dfrac{\partial H}{\partial \vec{p}}$ 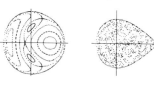	Poincaré Map

plate I [at the beginning of the book] for a set of interferometric pictures of a Bénard cell).

To describe the Bénard experiment theoretically, Lorenz truncated the complicated differential equations which describe this system (see Appendix A) and obtained the equations of the so-called Lorenz model:

$$\dot{X} = -\sigma X + \sigma Y \tag{1.4a}$$

$$\dot{Y} = rX - Y - XZ \tag{1.4b}$$

$$\dot{Z} = XY - bZ \tag{1.4c}$$

where σ and b are dimensionless constants which characterize the system, and r is the control parameter which is proportional to ΔT. The variable X is proportional to the circulatory fluid flow velocity, Y characterizes the temperature difference between ascending and descending fluid elements, and Z is proportional to the deviations of the vertical temperature profile from its equilibrium value.

A numerical analysis of this apparently simple set of nonlinear differential equations shows that its variables can exhibit chaotic motion above a threshold value r_c (see Appendix B). It should be noted, however, that the Lorenz equations describe the Bénard experiment only in the immediate vicinity of the transition from heat conduction to convection rolls because the spatial Fourier coefficients retained by Lorenz only describe simple rolls. The chaos found by Lorenz in eqns. (1.4 a–c) is, therefore, different from the chaos seen in the experimental power spectrum in Table 2. To describe the experimentally observed chaos, many more spatial Fourier components have to be retained.

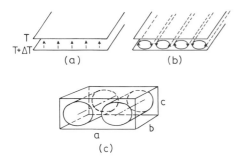

(a) (b)

(c)

Fig. 3: The Rayleigh-Bénard instability. a) and b) Transition from heat conduction to convection rolls in an infinitely extended two-dimensional fluid layer. c) Experiments to detect deterministic chaos in time are performed in a "match box".

Stirred Chemical Reactions

Another system in which chaotic motion has been studied experimentally in great detail is the Belousov-Zhabotinsky reaction. In this chemical process, an organic molecule (e. g. malonic acid) is oxidized by bromate ions; the oxidation is catalyzed by a redox system (Ce^{4-}/Ce^{3-}). The reactants, which undergo 18 elementary reaction steps (see Epstein et al., 1983), are:

$$Ce_2(SO_4)_3, \quad NaBrO_3, \quad CH_2(COOH)_2, \quad and \quad H_2SO_4 .$$

It is not our aim to describe these reactions in detail but to demonstrate that stirred chemical reactions provide convenient model systems to study the onset of chaos. Fig. 4 shows how a chemical reaction is maintained in a steady state away from equilibrium by continuously pumping the chemicals into a flow reactor where they are stirred to ensure spatial homogeneity. For example, the reaction

$$A + B \underset{k_2}{\overset{k_1}{\rightleftharpoons}} C \tag{1.5}$$

is described by the equations:

$$\dot{c}_A = -k_1 c_A c_B + k_2 c_C - r[c_A - c_A(0)] \qquad (1.6\,a)$$

$$\dot{c}_B = -k_1 c_A c_B + k_2 c_C - r[c_B - c_B(0)] \qquad (1.6\,b)$$

$$\dot{c}_C = k_1 c_A c_B - k_2 c_C - r \qquad (1.6\,c)$$

where eq. (1.6a) can be interpreted as follows. The concentration c_A decreases due to collisions between A and B (which generate C), increases due to decays of C (into A and B), and decreases if the flow rate r increases since for $k_1 = k_2 = 0$, eq. (1.6a) can be integrated to $c_A(t) - c_A(0) \sim \exp(-rt)$.

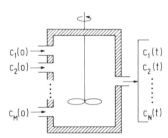

Fig. 4: Flow of chemicals in a well stirred reactor. $c_1(0) \ldots c_M(0)$ are the initial concentrations of the chemicals, and $c_1 \ldots c_N(t)$ are the output concentrations.

Generalizing, the reactions of M chemicals of concentrations c_i can be described by a set of first-order nonlinear differential equations

$$\dot{c}_i = g_i\{c_j\} - r[c_i - c_i(0)] \equiv F_i\{c_j, \lambda\} \qquad (1.7)$$

where the function $g_i\{c_j\}$ involves nonlinear terms of the form c_i^2 and $c_i c_j$ if three-body collisions are neglected. The reactions can be studied as a function of the set of control parameters $\lambda \equiv \{c_i(0), k_j, r\}$ that involves the initial concentrations $\{c_i(0)\}$, the temperature dependent reaction velocities $\{k_j\}$ and the flow rate r. Since r influences all individual reactions and can be easily manipulated by changing the pumping rate of the chemicals, it is usually used as the only control parameter.

Let us now come back to the Belousov Zhabotinsky reaction. The variable which signals chaotic behavior in this system is the concentration c of the Ce^{4+} ions. It is measured by the selective light absorption of these ions. The mean residence time of the substances in the open reactor (i.e. r^{-1}) acts as an external control parameter corresponding to ΔT in the previous experiment.

Table 2 shows a transition to chaos in this system which is detected via the change in the autocorrelation function

$$C(\tau) = \lim_{T \to \infty} \frac{1}{T} \int_0^T dt\, \hat{c}(t)\hat{c}(t+\tau); \quad \hat{c}(t) = c(t) - \lim_{T \to \infty} \frac{1}{T} \int_0^T dt\, c(t). \quad (1.8)$$

This function measures the correlation between subsequent signals. It remains constant or oscillates for regular motion and decays rapidly (mostly with an exponential tail) if the signals become uncorrelated in the chaotic regime (Roux et al., 1981).

It should be noted that the power spectrum $P(\omega)$ is proportional to the Fourier transformation of $C(\tau)$

$$P(\omega) = |\hat{c}(\omega)|^2 \propto \lim_{T \to \infty} \int_0^T d\tau \, e^{i\omega\tau} C(\tau) \tag{1.9}$$

that is, both quantities contain the same information. Eq. (1.9) can be derived by the usual rules for Fourier transformations if one continues $\hat{c}(t)$ periodically in T so that $\hat{c}(t) = \hat{c}(t + nT)$ and n is an integer which leads to

$$\hat{c}(\omega) = \lim_{T \to \infty} \int_0^T dt^{i\omega t} \hat{c}(t) = \lim_{T \to \infty} \int_{-T}^T dt \cos(\omega t)\hat{c}(t). \tag{1.10}$$

Hénon-Heiles System

Let us finally have a look at a simple nonintegrable example from classical mechanics that displays chaotic motion. In 1964 Hénon and Heiles numerically studied the canonical equations of motion of the Hamiltonian

$$H = \frac{1}{2}(p_1^2 + p_2^2) + \frac{1}{2}(q_1^2 + q_2^2) + q_1^2 q_2 - \frac{1}{3} q_2^3. \tag{1.11}$$

This equation describes, in Cartesian coordinates q_1 and q_2, two nonlinearly coupled harmonic oscillators and, in polar coordinates (r, θ), a single particle in a noncentrosymmetric potential

$$V(r, \theta) = \frac{r^2}{2} + \frac{r^3}{3} \sin(3\theta) \tag{1.12}$$

that is obtained from $1/2(q_1^2 + q_1^2) + q_1^2 q_2 - 1/3\, q_2^3$ via $q_1 = r \cos \theta$ and $q_2 = r \sin \theta$ (see Fig. 5).

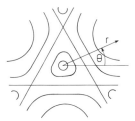

Fig. 5: Equipotential lines $V(r, \theta) =$ const. of the Hénon-Heiles system (eq. 1.12) in polar coordinates.

Their investigation was motivated by empirical evidence that a star moving in a weakly disturbed cylindrically symmetric potential should have, in addition to the energy E, another constant of the motion I. This would imply that, for bounded motion, the trajectory of the Hénon Heiles system in phase space

$$\vec{x}(t) = [p_1(t), p_2(t), q_1(t), q_2(t)] \tag{1.13}$$

where p_1, p_2 are the momenta, is confined (via $E[\vec{x}(t)] = $ const. and $I[\vec{x}(t)] = $ const.) to a two-dimensional closed surface. In order to check this proposal, Hénon and Heiles followed a method introduced by Poincaré (1893) and plotted the points in which the trajectory $\vec{x}(t)$ cuts the (p_2, q_2) plane. If the motion would be confined to a two-dimensional manifold, these points should form closed curves corresponding to the cut of the two-dimensional closed surface with the (p_2, q_2) plane. The last line in Table 2 shows that, at low energies, different initial conditions in the Hénon Heiles system indeed lead to closed curves in the Poincaré map. However for high enough energy (which acts as control parameter for this system) the lines decay, and the points in the Poincaré map of the Hénon Heiles model become plane-filling. This indicates, according to Fig. 6, highly irregular chaotic motion in phase space and the absence of an additional constant of the motion I.

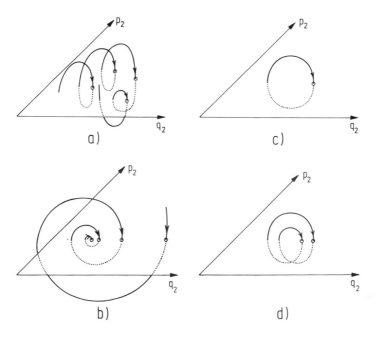

Fig. 6: Qualitatively different trajectories can be distinguished by their Poincaré sections: a) chaotic motion; b) approach of a fixed point; c) cycle; d) cycle of period two.

To summarize:

1. We have presented four possible criteria for chaotic motion:
   - The time dependence of the signal "looks chaotic".
   - The power spectrum exhibits broadband noise.
   - The autocorrelation function decays rapidly.
   - The Poincaré map shows space-filling points.

   In all four criteria, chaos is indicated by a qualitative change. Later, we will introduce some more quantitative measures to characterize deterministic chaos.

2. A common feature of the systems listed in Table 2 is that they can be characterized by low-dimensional first-order differential equations

$$\dot{\vec{x}} = \vec{F}(\vec{x}, \lambda); \quad \vec{x} = (x_1 \ldots x_d) \tag{1.14}$$

that are autonomous (i.e. $\vec{F}$ does not contain the time explicitly) and nonlinear ($\vec{F}$ is a nonlinear function of the $\{x_j\}$).

These equations lead to chaotic motion if an external control parameter λ (which can be the amplitude of the driving torque for the pendulum or the temperature difference ΔT in the Lorenz model, etc.) is varied. One distinguishes between *conservative systems,* for which a volume element in phase space $\{\vec{x}\}$ only changes its shape but retains its volume in the course of time (an example is the Hénon-Heiles Hamiltonian system for which the Liouville theorem holds) and *dissipative systems,* for which volume elements shrink as time increases (see also Chapter 6).

It is often convenient to study the flow described by the equations of motion (1.14) via the corresponding $(d - 1)$-dimensional Poincaré map

$$\vec{x}(n + 1) = \vec{G}[\vec{x}(n), \lambda]; \quad \vec{x}(n) = [x_1(n), \ldots x_{d-1}(n)] \tag{1.15}$$

that is generated by cutting the trajectory in d-dimensional phase space with a $(d - 1)$-dimensional hyperplane (see Fig 6) and by denoting the points which are generated with increasing time by $\vec{x}(1), \vec{x}(2) \ldots$ etc. The classifications "conservative" and "dissipative" can then be generalized from flows to maps (see Chapter 5, eqns. (5.6a, b)).

Let us finally comment on the way in which we shall proceed with our description of real physical systems. One can generally distinguish several levels of description as shown in Fig. 7.

A typical example of such a reduction process is given in Appendix A where the Navier-Stokes equations (which already represent a coarse grained description of molecular motion) are, for the boundary conditions of a Bénard experiment, reduced to the three differential equations of the Lorenz model which lead in turn to *different* Poincaré maps (see. Figs. 49, 67) corresponding to different parameter values. Another example has been given by Haken (1975) who reduced the quantum mechanical equations for a single mode laser to a system of three rate equations (which is equivalent

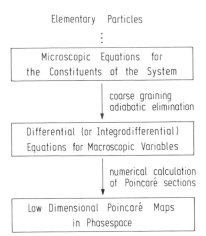

Fig. 7: Hierarchy for the levels of description of dynamical systems.

to the Lorenz system) by concentrating on macroscopic photon densities and using the adiabatic approximation ("Slaving principle").

In the following, we shall not be concerned with the details of this reduction process since the step from microscopic equations to differential equations for macroscopic variables has already been covered in several excellent books (Haken 1982, 1984), and the reduction of differential equations to Poincaré maps can be done numerically. It should also be clear that this reduction of a many-constituent system to a map, which describes only a few degrees of freedom, is not always possible; a counterexample would be fully developed spatio-temporal turbulence. Nevertheless, since it has been found experimentally effective for many physical systems (see the following chapters), we shall in the remainder of this book concentrate mostly on the last level in Fig. 7 where the dynamics of a system has been reduced to a one- or two-dimensional Poincaré map. We shall use these maps as starting points for our description of chaotic systems in the same sense as one uses the (coarse grained) Ginzburg-Landau Hamiltonian to derive universal properties of second-order phase transitions (Wilson and Kogut 1974). It will then be shown that only some general features of these maps (such as, for example, the existence of a simple maximum) determine how chaos emerges. The various "routes to chaos" differ in the way in which the signal behaves before becoming completely chaotic.

Although universal features of several routes to chaos have been discovered and verified experimentally it should be stated explicitly that it is presently practically impossible to theoretically predict, for example, from the Navier Stokes equations with given boundary conditions the route to chaos for a given experimental hydrodynamic system. This situation could be compared to ordinary second-order phase transitions where one knows a lot about universality classes and critical exponents (for example, of magnetic systems) but where it is still a formidable and often unsolved problem to predict the transition temperature of a given magnet (Ma 1976). However, this limitation should not disappoint us. The beauty of physics reveals itself only after asking the right questions, and it seems, from the results summarized in this book (see especially table 12 on page 182) that it is equally so for dynamical systems where the question about universal features has led to the discovery of a beautiful unifying pattern behind different phenomena in this field.

1.2   The Periodically Kicked Rotator

One of the simplest dynamical systems which displays chaotic behavior in time is the periodically kicked damped rotator shown in Fig 8.

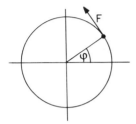

Fig. 8: Rotator kicked by a force F.

Its equation of motion is

$$\ddot{\varphi} + \Gamma\dot{\varphi} = F \equiv Kf(\varphi) \sum_{n=0}^{\infty} \delta(t - nT), \quad n \text{ integer} \tag{1.16}$$

where the dot denotes the time derivatives; Γ is the damping constant, T is the period between two kicks, and we normalize the moment of inertia to unity. If we make the substitutions $x = \varphi$, $y = \dot{\varphi}$, $z = t$, equation (1.16) can be rewritten as a system of first-order nonlinear autonomous differential equations

$$\dot{x} = y \tag{1.17a}$$
$$\dot{y} = -\Gamma y + Kf(x) \sum_{n=0}^{\infty} \delta(z - nT) \tag{1.17b}$$
$$\dot{z} = 1 . \tag{1.17c}$$

These can be reduced to a two-dimensional map for the variables $(x_n, y_n) = \lim_{\varepsilon \to 0} [x(nT - \varepsilon), y(nT - \varepsilon)]$ by integration. The general solution of (1.17b) for $(n + 1)T - \varepsilon > t > nT - \varepsilon$ is

$$y(t) = y_n e^{-\Gamma(t-nT)} + K \sum_{m=0}^{\infty} f(x_m) \int_{nT-\varepsilon}^{t} dt' e^{\Gamma(t'-t)} \delta(t' - mT) . \tag{1.17}$$

This yields

$$y_{n+1} = e^{-\Gamma T}[y_n + Kf(x_n)] \tag{1.18a}$$

and by integrating (1.17a) using (1.18a) we obtain:

$$x_{n+1} = x_n + \frac{1 - e^{-\Gamma T}}{\Gamma}[y_n + Kf(x_n)] . \tag{1.18b}$$

Equations (1.12a) and (1.12b) are the main results of this section. They reduce the initial set of three-dimensional differential equations to a two-dimensional discrete map, which yields a stroboscopic picture of the variables. Below, we list several important limits of this two-dimensional map which will be discussed in some detail in the following sections.

Logistic Map

This is a one-dimensional quadratic map defined by

$$x_{n+1} = rx_n(1 - x_n) \tag{1.19}$$

where r is an external parameter, and the range of x_n is changed from a circle to the interval [0,1]. It can be obtained from (1.18b) in the strong damping limit ($\Gamma \rightarrow \infty$) if $K \rightarrow \infty$, so that $\Gamma/K = 1$ and $f(x_n) = (r-1)x_n - rx_n^2$.

Hénon Map

This can be considered as a two-dimensional extension of the logistic map (Hénon, 1976):

$$x_{n+1} = 1 - ax_n^2 + y_n \tag{1.20a}$$
$$y_{n+1} = bx_n \tag{1.20b}$$

where a and $|b| \leq 1$ are external parameters.
To obtain this map from (1.18a–b), we rewrite these equations as

$$y_{n+1} = e^{-\Gamma T}[y_n + Kf(x_n)] \tag{1.21a}$$

$$x_{n+1} = x_n + \frac{e^{\Gamma T} - 1}{\Gamma}y_{n+1} \tag{1.21b}$$

and solve (1.21 b) for y_{n+1}:

$$y_{n+1} = (x_{n+1} - x_n)\,\Gamma/(e^{\Gamma T} - 1)\,. \tag{1.22}$$

If we put y_{n+1} and y_n back into (1.21 a), this becomes for $T = 1$:

$$x_{n+1} + e^{-\Gamma} x_{n-1} = (1 + e^{-\Gamma})\,x_n + \frac{1 - e^{-\Gamma}}{\Gamma}\,Kf(x_n)\,. \tag{1.23}$$

Choosing

$$\frac{1 - e^{-\Gamma}}{\Gamma}\,Kf(x_n) \equiv -(1 + e^{-\Gamma})\,x_n - 1 + ax_n^2\,; \quad b \equiv -e^{-\Gamma} \tag{1.24}$$

equation (1.23) yields

$$x_{n+1} = 1 - ax_n^2 + bx_{n-1} \tag{1.25}$$

which is equivalent to (1.20 a–b). (Our derivation holds only for $b < 0$, but the map is mathematically defined for $-1 \le b \le 1$.)

Chirikov Map

This is simply the map of an undamped ($\Gamma \to 0$) rotator that is kicked by an external force $Kf(x_n) = -K \sin x_n$ (Chirikov, 1979). In this limit eq. (1.18 a–b) reduce to

$$p_{n+1} = p_n - K \sin \theta_n \tag{1.26a}$$
$$\theta_{n+1} = \theta_n + p_{n+1} \tag{1.26b}$$

where we have chosen $T = 1$ and introduced the conventional notation $x_n = \theta_n$ and $y_n = p_n$.

We shall see in the following chapters that despite the apparent simplicity of all three maps, their iterates exhibit extremely rich and physically interesting structures.

2 Piecewise Linear Maps and Deterministic Chaos

The nonlinear Poincaré maps introduced in the previous chapter still lead to a rather complicated dynamical behavior (as we shall see in Chapter 3). In this section, we therefore study some simple one-dimensional piecewise linear maps. Although these maps are not directly connected to physical systems, they are extremely useful models which, in part one of this section, allow us to explain the mechanism which leads to deterministic chaos. In the second part, we will introduce three quantitative measures which characterize chaotic behavior and calculate these quantities explicitly for a triangular map. Finally, in Section 3 we show that the iterates of certain one-dimensional maps can display deterministic diffusion.

2.1 The Bernoulli Shift

Let us consider the one-dimensional map

$$x_{n+1} = \sigma(x_n) \equiv 2x_n \bmod 1; \quad n = 0, 1, 2 \ldots \tag{2.1}$$

which is shown in Fig. 9.

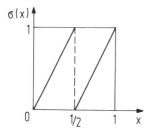

Fig. 9: The transformation $\sigma(x) = 2x \bmod 1$.

If we start with a value x_0, the map generates a sequence of iterates x_0, $x_1 = \sigma(x_0)$, $x_2 = \sigma(x_1) = \sigma(\sigma(x_0))$ ... In order to investigate the properties of this sequence we write x_0 in binary representation:

$$x_0 = \sum_{v=1}^{\infty} a_v 2^{-v} \triangleq (0, a_1 a_2 a_3 \ldots) \tag{2.2}$$

where a_v has the values zero or unity. For $x_0 < 1/2$, we have $a_1 = 0$, and $x_0 > 1/2$ implies $a_1 = 1$. Therefore, the first iterate $\sigma(x_0)$ can be written as

$$\sigma(x_0) = \begin{cases} 2x_0 & \text{for} \quad a_1 = 0 \searrow \\ 2x_0 - 1 & \text{for} \quad a_1 = 1 \nearrow \end{cases} = (0, a_2 a_3 a_4 \ldots) \tag{2.3}$$

i.e. the action of σ on the binary representation of x is to delete the first digit and shift the remaining sequence to the left. This is called the *Bernoulli shift*.

The Bernoulli property of $\sigma(x)$ demonstrates:

1. The sensitive dependence of the iterates of σ on the initial conditions. Even if two points x and x' differ only after their nth digit a_n, this difference becomes amplified under the action of σ, and their nth iterates $\sigma^n(x)$ and $\sigma^n(x')$ already differ in the first digit because $\sigma^n(x) = (0, a_n \ldots)$ where $\sigma^2(x) \equiv \sigma[\sigma(x)]$, etc.

2. The sequence of iterates $\sigma^n(x_0)$ has the same random properties as successive tosses of a coin. To see this, we attach to $\sigma^n(x_0)$ the symbol R or L depending on whether the iterate is contained in the right or left part if the unit interval. If we now prescribe an arbitrary sequence $R L L R \ldots$, e.g. by tossing a coin, we can always find an x_0 for which the series of iterates x_0, $\sigma^1(x_0)$, $\sigma^2(x_0) \ldots$ generates this sequence. This follows because $\sigma^n(x_0) = (0, a_n a_{n+1} \ldots)$ corresponds to R or L if and only if $a_n = 1$ or $a_n = 0$; i.e., the sequence $R L L R \ldots$ is isomorphous to the binary representation of x_0

$$
\begin{array}{cccccc}
x_0 = (0, & 1 & 0 & 0 & 1 \ldots) \\
 & \updownarrow & \updownarrow & \updownarrow & \updownarrow \\
 & R & L & L & R
\end{array} \tag{2.4}
$$

Thus, the prescription of a sequence by tossing a coin becomes equivalent to choosing a special value of x_0.

3. The mechanism by which ergodicity emerges in a deterministic system. Let us first note that we can approximate each point x in the unit interval arbitrarily well by a finite sequence of binary digits $0, a_1 a_2 \ldots a_n$ up to a difference $\varepsilon = 2^{-n}$, say. It will now be shown that the images $\sigma^r(x_0)$ $(r = 1, 2, 3 \ldots)$ of an "arbitrary" irrational number $x_0 \in [0, 1]$ approach x to an order ε an infinite number of times; i.e., the system behaves ergodically.

This follows because a) almost all irrational numbers in [0, 1] (with the exception of a set of measure zero) in their binary representation contain any finite sequence of digits infinitely often (see References on page 249) and b) the Bernoulli property of $\sigma(x)$ shifts these sequences to the initial position as depicted in Fig. 10. This argument goes right to the heart of the problem of chaotic motion in deterministic systems, and it shows how chaos arises from the amplification of the intrinsic "numerical noise" of irrational numbers.

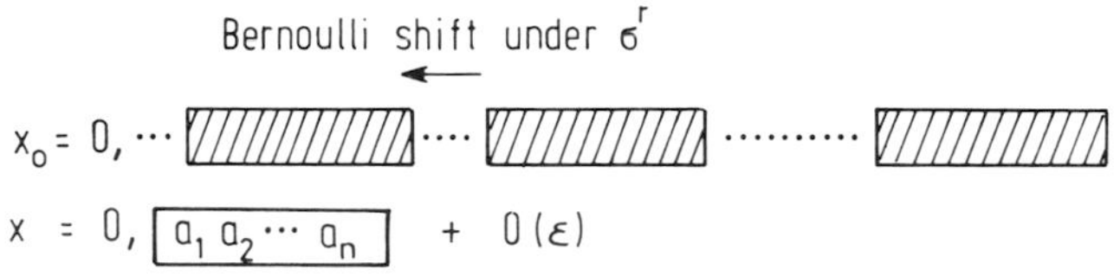

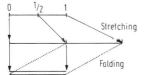

Bernoulli shift under σ^r

$x_0 = 0, \cdots$

$x = 0, \boxed{a_1 \, a_2 \cdots a_n} \quad + \quad 0(\varepsilon)$

Fig. 10: Emergence of ergodicity by a Bernoulliy shift in irrational numbers.

This mechanism of generating deterministic chaos is also quite universal. Its two basic ingredients are the stretching and backfolding property of the map.

Initially, for $x_0 < 1/2$ say, x_0 becomes stretched after each iteration by a factor 2 (see Fig. 11). But for $n > n_0$ with $2^{n_0} \cdot x_0 \geq 1$, the second branch of $\sigma(x)$ becomes important, and x_n is folded back to the unit interval as shown in Fig. 11.

0   1/2   1

Stretching

Folding

Fig. 11: Stretching and folding of the unit interval under the action of $\sigma(x)$.

For a general nonlinear map of the unit interval onto itself, the combination of stretching and backfolding (due to the restriction to [0, 1]) drives the iterates of an initial point repeatedly over the unit interval and leads to chaotic motion.

Let us briefly comment on the possible physical consequences of this stretching property of nonlinear maps. The initial conditions (i.e. the x_0) of a physical system can only be determined with finite precision. This "arbitrarily" small but finite error becomes exponentially amplified ($\sigma^n(x_0) = 2^n x_0 \bmod 1$) via the nonlinear evolution equation. Such an equation thus acts like a microscope which makes the limits of our precision in physical measurements visible. Can we, therefore, anticipate that the concept of the continuum with its distinction of rational and irrational numbers is nonphysical and all physical variables will be quantized? (The Heisenberg uncertainty relation, which limits the precision of our observations for conjugate variables, has also been found in a gedanken experiment in which one tries to measure the location and the momentum of an electron via a light microscope with arbitrary accuracy.) This and related questions, and speculations, are discussed in an interesting article by J. Ford in "Physics Today", April 1983.

2.2 Characterization of Chaotic Motion

In this section, we introduce the Liapunov exponent as well as the invariant measure and the correlation function as quantitative measures to characterize the chaotic motion which is generated by one-dimensional Poincaré maps.

Liapunov Exponent

We have already seen in the previous section that adjacent points become separated under the action of a map

$$x_{n+1} = f(x_n) \tag{2.5}$$

which leads to chaotic motion. The Liapunov exponent $\lambda(x_0)$ measures this exponential separation as shown in Fig. 12.

Fig. 12: Definition of the Liapunov exponent.

From Fig. 12 one obtains:

$$\varepsilon\, e^{N\lambda(x_0)} = |f^N(x_0 + \varepsilon) - f^N(x_0)| \tag{2.6}$$

which, in the limits $\varepsilon \to 0$ and $N \to \infty$, leads to the correct formal expression for $\lambda(x_0)$:

$$\lambda(x_0) = \lim_{N\to\infty} \lim_{\varepsilon\to 0} \frac{1}{N} \log \left| \frac{f^N(x_0 + \varepsilon) - f^N(x_0)}{\varepsilon} \right| \tag{2.7}$$

$$= \lim_{N\to\infty} \frac{1}{N} \log \left| \frac{df^N(x_0)}{dx_0} \right|.$$

This means that $e^{\lambda(x_0)}$ is the average factor by which the distance between closely adjacent points becomes stretched after one iteration.

The Liapunov exponent also measures the average loss of information (about the position of a point in [0, 1]) after one iteration. In order to see this, we use in (2.7) the chain rule

$$\frac{d}{dx} f^2(x) \Big|_{x_0} = \frac{d}{dx} f[f(x)] \Big|_{x_0} = f'[f(x_0)] f'(x_0) \tag{2.8}$$

$$= f'(x_1) f'(x_0); \quad x_1 \equiv f(x_0)$$

to write the Liapunov exponent as

$$\lambda(x_0) = \lim_{N \to \infty} \frac{1}{N} \log \left| \frac{d}{dx_0} f^N(x_0) \right| = \lim_{N \to \infty} \frac{1}{N} \log \left| \prod_{i=0}^{N-1} f'(x_i) \right|$$

$$= \lim_{N \to \infty} \frac{1}{N} \sum_{i=0}^{N-1} \log |f'(x_i)| \ . \tag{2.9}$$

As a next step, we discuss the loss of information after one iteration with a linear map. We separate [0, 1] into n equal intervals and assume that a point x_0 can occur in each of them with equal probability $1/n$. By learning which interval contains x_0, we gain the information

$$I_0 = - \sum_{i=1}^{n} \frac{1}{n} \operatorname{ld} \frac{1}{n} = \operatorname{ld} n \tag{2.10}$$

where ld is the logarithm to the base 2 (see Appendix F). If we decrease n, the information I_0 is reduced, and it becomes zero for $n = 1$.

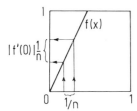

Fig. 13: Increase of an interval $1/n$ by a linear map.

It is shown in Fig. 13 that a linear map $f(x)$ changes the length of an interval by a factor $a = |f'(0)|$. The corresponding decrease of resolution leads to a loss of information after the mapping:

$$\Delta I = - \sum_{i=1}^{\frac{n}{a}} \frac{a}{n} \operatorname{ld} \frac{a}{n} + \sum_{i=1}^{n} \frac{1}{n} \operatorname{ld} \frac{1}{n} = - \operatorname{ld} a = - \operatorname{ld} |f'(0)| \tag{2.11}$$

Generalizing this expression to a situation where $|f'(x)|$ varies from point to point and averaging over many iterations lead to the following expression for the mean loss of information:

$$\overline{\Delta I} = - \lim_{N \to \infty} \frac{1}{N} \sum_{i=0}^{N-1} \operatorname{ld} |f'(x_i)| \tag{2.12}$$

which is, via (2.9), proportional to the Liapunov exponent:

$$\lambda\,(x_0)\;=\;(\log 2)\;\cdot\;|\overline{\Delta I}|\;.\tag{2.13}$$

This relation between the Liapunov exponent and the loss of information is a first step towards characterization of chaos in a coordinate-invariant way, as will be explored on a deeper level in Chapter 5.

By way of an example, we now calculate the Liapunov exponent for the triangular map,

$$\Delta\,(x)\;=\;r\left(1\,-\,2\,\left|\frac{1}{2}\,-\,x\right|\right)\tag{2.14}$$

shown in Fig. 14.

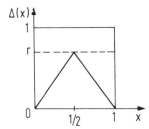

Fig. 14: The triangular map $\Delta\,(x)$.

The function $\Delta\,(x)$ serves as a useful model because, for $r > 1/2$, it generates chaotic sequences x_0, $\Delta\,(x_0)$, $\Delta\,[\Delta\,(x_0]\,\ldots$, and due to its simple form, all quantities that characterize the chaotic state can be calculated explicitly.

In order to get acquainted with this map, we first consider its fixed points and their stability for different values of r.

Generally, a point x^* is called a fixed point of a map $f(x)$ if

$$x^*\;=\;f(x^*)\tag{2.15}$$

i.e. the fixed points are the intersections of $f(x)$ with the bisector.

A fixed point is locally stable if all points x_0 in the vicinity of x^* are attracted to it, i.e., if the sequence of iterates of x_0,

$$x_0,\,x_1,\,x_2\,..\,x_n,\,\ldots\;\equiv\;x_0,\,f(x_0),\,f[f(x_0)]\,\ldots\underbrace{f[f\,\ldots\,f(x_0)\,..],}_{n}\,\ldots\tag{2.16}$$

converges to x^*. The analytical criterion for local stability is

$$\left|\frac{\mathrm{d}}{\mathrm{d}x^*}\,f(x^*)\right|\;<\;1\tag{2.17}$$

because the distance $\delta_n \, to \, x^*$ shrinks as

$$\delta_{n+1} = |x_{n+1} - x^*| = |f(x_n) - x^*|$$

$$= |f(x^* + \delta_n) - x^*| \simeq \left| \frac{\mathrm{d}}{\mathrm{d}x^*} f(x^*) \right| \cdot \delta_n \tag{2.18}$$

Fig. 15 a shows that for $r < 1/2$ the origin $x = 0$ is the only stable fixed point to which all points [0, 1] are attracted.

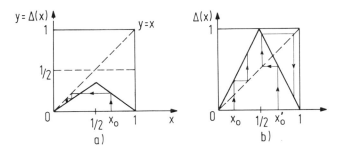

Fig. 15: a) Stable fixed point at $x^* = 0$ for $r < 1/2$; b) two unstable fixed points for $r = 1$.

For $r > 1/2$ two unstable fixed points emerge. Fig. 15 b shows how, for $r = 1$, the iterates of x_0 and x_0' move away from the "fixed points" $x_1 = 0$ and $x_2 = 2/3$, respectively. In the following, we shall consider only the case $r = 1$, which is respresentative for $r > 1/2$.

What can we say about a sequence of iterates if there are no stable fixed points? First of all we notice that points, which are close together, become more and more separated during the first iterations, as shown in Fig. 16.

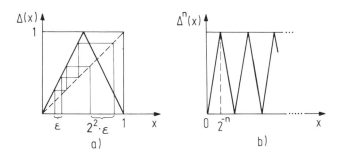

Fig. 16: a) Separation of points by iteration with $\Delta(x)$ and b) the nth iterate $\Delta^n(x)$.

If we plot the nth iterate $\Delta^n(x)$, we see from Fig. 16 that again it is piecewise linear and has the slope $\left| \frac{\mathrm{d}}{\mathrm{d}x} \Delta^n(x) \right| = 2^n$, except for the countable set of points $j \cdot 2^{-n}$

where $j = 0, 1 \dots 2^n$. Therefore, the separation of "almost all" points x_0, $x_0 + \varepsilon$ grows exponentially with n after n iterations, and the Liapunov exponent becomes (independent of x_0):

$$\lambda = \log 2 . \tag{2.19}$$

For the general triangular map (2.14), the Liapunov exponent simply becomes $\lambda = \log 2 r$, and for $r > 1/2$ we have $\lambda > 0$; i.e., we lose information about the position of a point in [0, 1] after an iteration, whereas $r < 1/2$ implies $\lambda < 0$, and we gain information because all points are attracted to $x^* = 0$.

The Liapunov exponent changes sign at $r = 1/2$ and, therefore, acts like an "order parameter", which indicates the onset of chaos, as shown in Fig. 17.

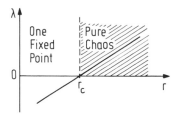

Fig. 17: The Liapunov exponent for the triangular map as a function of r in the vicinity of r_c.

To make the analogy to critical phenomena even closer, we observe that $\lambda = \log 2 r$ scales with a power law near the "critical point" $r_c = 1/2$.

$$\lambda \propto (r - r_c) . \tag{2.20}$$

This shows that even the simple transition to chaos in the triangular map displays some features that are remiscent of an equilibrium phase transition. As we have already mentioned before, we will investigate this aspect more generally in Chapter 3. It should also be noted here that the definition of the Liapunov exponent can be extended to higher dimensional maps. This will be treated in Chapter 5, where we will also discuss the relation between the Liapunov exponent and the Kolmogorov entropy and its possible connection to the Hausdorff dimension. But before we come to these problems, we will first investigate the question of how the iterates of a one-dimensional map are distributed over the unit interval.

Invariant Measure

The invariant measure $\rho(x)$ determines the density of the iterates of a unimodular map

$$x_{n+1} = f(x_n), \quad x_n \varepsilon [0, 1], \quad n = 0, 1, 2, .. \tag{2.21}$$

over the unit interval and is defined via

$$\rho(x) \equiv \lim_{N \to \infty} \frac{1}{N} \sum_{i=0}^{N} \delta[x - f^i(x_0)] . \tag{2.22}$$

If $\rho(x)$ does not depend on x_0, the system is called ergodic (see also page 202). For this case, eq. (2.22) allows us to write "time averages" over a function $g(x)$ as averages over the invariant measure.

$$\lim_{N \to \infty} \frac{1}{N} \sum_{i=0}^{N} g(x_i) \equiv \lim_{N \to \infty} \frac{1}{N} \sum_{i=0}^{N} g[f^i(x_0)] = \int_0^1 dx \rho(x) g(x) . \tag{2.23}$$

This is the one-dimensional analog of the thermodynamic average in classical statistical mechanics which allow us (if the motion in phase space is ergodic) to replace the time average by an ensemble average over a stationary distribution $\tilde{\rho}$:

$$\lim_{T \to \infty} \frac{1}{T} \int_0^T dt A[\vec{x}(t)] = \int d\vec{x} \tilde{\rho}(\vec{x}) A(\vec{x}) . \tag{2.24}$$

Here A is a function of the time-dependent vector $\vec{x} = [\vec{p}(t), \vec{q}(t)]$ which is composed of the coordinates $\vec{q}$ and momenta $\vec{p}$ which follow Hamilton's equations,

$$\dot{q}_i = \frac{\partial H}{\partial p_i}, \quad \dot{p}_i = -\frac{\partial H}{\partial q_i} \tag{2.25}$$

and $\tilde{\rho}$ is, for example, the microcanonical distribution $\tilde{\rho} = \delta[H(\vec{x}) - E]$ for an isolated system of energy E. Note, however, that our one-dimensional example corresponds to a dissipative system (see e. g. Chapter 1, eq. (1.15)) whereas Hamilton's equations (2.25) describe a conservative model.

For Hamiltonian systems, the dynamical behavior of a general density distribution $\rho(\vec{x}, t)$ in phase space is described by Liouvilles's equation:

$$\dot{\rho}(\vec{x}, t) = -i L \rho(\vec{x}, t) \tag{2.26}$$

where

$$L = i \left[\frac{\partial H}{\partial \vec{p}} \frac{\partial}{\partial \vec{q}} - \frac{\partial H}{\partial \vec{q}} \frac{\partial}{\partial \vec{p}} \right] \tag{2.27}$$

is the Liouville operator.

The corresponding evolution equation for our one-dimensional model whose time evolution is given by the map (2.21) can be derived as follows. If we have a point x_0, it evolves to $f(x_0)$ after one iteration. This means that a delta-function distribution $\delta(x - x_0)$ evolves after one time step to $\delta[x - f(x_0)]$ which can be written as

$$\delta[x - f(x_0)] = \int_0^1 dy \delta[x - f(y)] \delta(y - x_0). \tag{2.28}$$

Generalizing this to the evolution of an arbitrary density $\rho_n(x)$ at time n we obtain the so-called Frobenius-Perron equation

$$\rho_{n+1}(x) = \int_0^1 dy\, \delta[x - f(y)]\, \rho_n(y) \tag{2.29}$$

which governs the time evolution of $\rho_n(x)$. The invariant measure $\rho(x)$ has to be stationary because eq. (2.23) makes sense only if $\rho(x)$ is independent of time n, that is, $\rho(x)$ is an eigenfunction of the Frobenius-Perron operator with eigenvalue 1:

$$\rho(x) = \int_0^1 dy\, \delta[x - f(y)]\, \rho(y). \tag{2.30}$$

Formally, this equation has many solutions (e.g. $\delta(x - x^*)$ where $x^* = f(x^*)$ is an unstable fixed point). But fortunately, only one of the solutions is physically relevant, namely that one which is, for example, obtained by solving eq. (2.30) on a computer. In the presence of weak random noise (which is caused by rounding errors in the computer or physical fluctuations in real systems), the probability to hit an unstable repelling fixed point x^* is zero, and therefore such spurious solutions are automatically eliminated (Eckmann and Ruelle, 1985). In the following, the invariant measure $\rho(x)$, always means the physically relevant invariant measure which is stable if a small random noise is added to the system.

Let us consider again, as an example, the triangular map at $r = 1$:

$$\Delta(x) = \begin{cases} 2x & \text{for } x \le \dfrac{1}{2} \\[2ex] 2(1 - x) & \text{for } x > \dfrac{1}{2} \end{cases} \tag{2.31}$$

In this case, eq. (2.30) becomes:

$$\rho(x) = \frac{1}{2}\left[\rho\left(\frac{x}{2}\right) + \rho\left(1 - \frac{x}{2}\right)\right] \tag{2.32}$$

which has the obvious normalized solution $\rho(x) = 1$.

We can also show that this solution is unique. Starting from an arbitrary normalized distribution $\rho_0(x)$, and operating on it n times with (2.29), yields

$$\rho_n(x) = \frac{1}{2^n}\sum_{j=1}^{2^{n-1}}\left[\rho_0\left(\frac{j-1}{2^{n-1}} + \frac{x}{2^n}\right) + \rho_0\left(\frac{j}{2^{n-1}} - \frac{x}{2^n}\right)\right] \tag{2.33}$$

which converges towards

$$\rho(x) = \lim_{n \to \infty}\rho_n(x) = \frac{1}{2}\left[\int_0^1 dx\rho_0(x) + \int_0^1 dx\rho_0(x)\right] = 1. \tag{2.34}$$

This means that, for the triangular map at $r = 1$, the chaotic sequence of iterates x_0, $f(x_0), f(f((x_0))) \ldots$ uniformly covers the interval $[0, 1]$, and the system is ergodic. As in the case of the Liapunov exponent, we will later study invariant density for more complicated maps and show that it is not always a constant.

Correlation Function

The correlation function $C(m)$ for a map (2.21) is defined by

$$C(m) = \lim_{N \to \infty} \frac{1}{N} \sum_{i=0}^{N-1} \hat{x}_{i+m} \hat{x}_i \tag{2.35}$$

where

$$\hat{x}_i = f^i(x_0) - \bar{x}; \quad \bar{x} = \lim_{N \to \infty} \frac{1}{N} \sum_{i=0}^{N-1} f^i(x_0). \tag{2.36}$$

From this definition follows that $C(m)$ yields another measure for the irregularity of the sequence of iterates $x_0, f(x_0), f^2(x_0) \ldots$ It tells us, how much the deviations, of the iterates from their average value,

$$\hat{x}_i = x_i - \bar{x} \tag{2.37}$$

that are m steps apart (i. e. $\hat{x}_{i+m}$ and $\hat{x}_i$) know about each other, on the average.

If the invariant measure $\rho(x)$ for a given map $f(x)$ is known, $C(m)$ can be written in the following form:

$$C(m) = \int_0^1 dx \rho(x) x f^m(x) - \left[\int_0^1 dx \rho(x) x \right]^2. \tag{2.38}$$

Here, we used the commutative property of the iterates,

$$x_{i+m} = f^{i+m}(x_0) = f^i f^m(x_0) = f^m f^i(x_0). \tag{2.39}$$

We, therefore, find for the example of the triangular map:

$$C(m) = \int_0^1 dx \, x \Delta^m(x) - \left[\int_0^1 dx \, x \right]^2 \tag{2.40a}$$

$$= \int_{-1/2}^{1/2} dy \, y \Delta^m\left(y + \frac{1}{2}\right) + \frac{1}{2} \int_{-1/2}^{1/2} dy \, \Delta^m\left(y + \frac{1}{2}\right) - \frac{1}{4}$$

$$= \frac{1}{12} \delta_{m,0} \tag{2.40b}$$

i. e. the sequence of iterates is delta-correlated.

This result follows because a) $\Delta^n(y + 1/2)$ is symmetric about $y = 0$; therefore, the first integral in (2.40b) vanishes for $m > 0$, and b) the second integral is independent of m, as shown in Fig. 18.

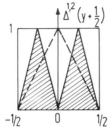

Fig. 18: The first and second iterates $\Delta^{1,2}(y + 1/2)$ are symmetric about $y = 0$; the triangular areas are independent of $m = 1, 2$.

To summarize:

We have found for a general one-dimensional map that a sequence $x_0, f(x_0) \ldots f^n(x_0) \ldots$ can be characterized a) by a Liapunov exponent, which tells us how adjacent points become separated under the action of f; b) by the invariant density, which serves as a measure of how the iterates become distributed over the unit interval; and c) by the correlation function $C(m)$, which measures the correlation between iterates that are m steps apart.

For the triangular map, the Liapunow exponent is $\lambda = \log 2r$, which changes its sign at $r = 1/2$. It, therefore, serves as an order parameter for the onset of chaos. For $r = 1$, the chaotic state is characterized by a constant stationary density $\rho(x) = 1$ and delta-correlated iterates, i.e. $C(m) = (1/12)\,\delta_{m,0}$.

2.3   Deterministic Diffusion

In this section, we show that the iterates of certain one-dimensional periodic maps diffuse. This diffusion indicates that the reduced map generates chaotic motion.

One normally associates diffusion with the Brownian motion of a particle in a liquid. Its equation of motion, in the case of high friction where the acceleration term $\propto \ddot{x}$ can be neglected, is

$$\dot{x} \propto \xi(t) . \tag{2.41}$$

The $\xi(t)$ are random forces, which are generated by the thermal agitation of the molecules. If one assumes, as usual, that the $\xi(t)$ are Gaussian-correlated,

$$\langle \xi(t) \rangle = 0; \quad \langle \xi(t)\,\xi(t') \rangle \propto \delta(t - t') \tag{2.42}$$

one obtains from eqns. (2.41) and (2.42):

$$\langle x(t) \rangle = 0 \quad \text{and} \quad \langle x^2(t) \rangle \propto t . \tag{2.43}$$

This means that the squared distance from the origin increases linearly with time if the particle is kicked by random forces (in contrast to $x^2 \propto t^2$ for a constant force $k \propto \dot{x}$). One can show, with a little more effort, that (2.43) also remains valid (for $t \to \infty$) if the acceleration term is retained (see, for example, Haken's book on *Synergetics* (1982)).

Let us now have a look at the piecewise linear periodic map

$$x_{\tau+1} = F(x_\tau) = x_\tau + f(x_\tau) ; \quad \tau = 0, 1, 2, \dots \tag{2.44}$$

where $f(x_\tau)$ is periodic in x_τ,

$$f(x_\tau + n) = f(x_\tau) , \quad n = 0, \pm 1, \pm 2 \dots , \tag{2.45}$$

shown in Fig. 19.

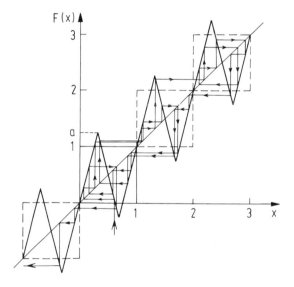

Fig. 19: Piecewise linear periodic map with a diffusive trajectory (after Grossmann, 1982).

One sees that the trajectory moves slowly away from the origin. Now we will show that this motion is in fact diffusive. However, this diffusion is not generated by random forces (as in the case of Brownian motion discussed above), but rather because the trajectory loses its "memory" within one or several boxes due to chaotic motion. To substantiate this statement, we calculate $\langle x^2 \rangle$ explicitly for the map (2.44).

We decompose the coordinate of a trajectory into the box number N_τ and the position $y_\tau \in [0, 1]$ within a box (Grossmann, 1982):

$$x_\tau = N_\tau + y_\tau . \tag{2.46}$$

The map (2.44) then becomes

$$N_{\tau+1} + y_{\tau+1} = F(N_\tau + y_\tau) = N_\tau + y_\tau + f(y_\tau) \tag{2.47}$$

which is equivalent to the coupled dynamical laws:

$$N_{\tau+1} - N_\tau = [y_\tau + f(y_\tau)] \equiv \Delta(y_\tau) \tag{2.48a}$$

$$y_{\tau+1} = y_\tau + f(y_\tau) - [y_\tau + f(y_\tau)] \equiv g(y_\tau) \tag{2.48b}$$

where $[z]$ denotes the integer part of z. Fig. 20 shows the function $\Delta(y_\tau)$, which is an integer number, describing the magnitude of the jump, and $g(x_\tau)$ gives the remaining part of the coordinate at $\tau + 1$.

Using (2.48a), the distances to the origin can be written as

$$N_t = \sum_{\tau=0}^{t-1} (N_{\tau+1} - N_\tau) = \sum_{\tau=0}^{t-1} \Delta(y_\tau) , \quad \text{for} \quad N_0 = 0 . \tag{2.49}$$

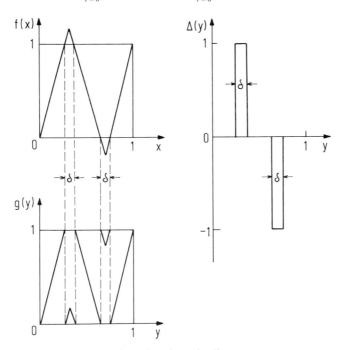

Fig. 20: Decomposition of a piecewise linear map.

This yields for the mean squared distance:

$$\langle N_t^2 \rangle = \sum_{\tau, \lambda}^{t-1} \langle \varDelta(y_\tau) \varDelta(y_\lambda) \rangle \tag{2.50}$$

where the average $\langle \dots \rangle$ is over all initial conditions y_0, and we assumed for simplicity $\langle N_t \rangle = 0$. For the case that the motion generated by $g(y)$ is so chaotic that there are no correlations among the y_τ, i.e.

$$\langle \varDelta(y_\lambda) \varDelta(y_\tau) \rangle \propto \delta_{\lambda, \tau} \tag{2.51}$$

one finds from (2.47):

$$\lim_{t \to \infty} \frac{\langle N_t^2 \rangle}{t} = \lim_{t \to \infty} \frac{1}{t} \sum_{\tau=0}^{t-1} \langle \varDelta^2(y_\tau) \rangle \tag{2.52}$$

$$= \int dy \rho(y) \varDelta^2(y) . \tag{2.53}$$

The step from (2.52) to (2.53) is only possible is $g(y)$ has an invariant density that obeys

$$\rho(y) = \int dx \delta [g(x) - y] \rho(x) . \tag{2.54}$$

Eq. (2.53) means that $\langle N_t^2 \rangle$ increases linearly with t, i.e.

$$\langle N_t^2 \rangle = 2Dt \quad \text{for} \quad t \gg 1 \tag{2.55}$$

with a diffusion coefficient

$$D \equiv \frac{1}{2} \int dy \rho(y) \varDelta^2(y) . \tag{2.56}$$

It should be clear from the derivation that diffusion occurs as long as the y_τ's are sufficiently uncorrelated such that the two sums in (2.50) contract to a single sum. (For completely correlated motion of the y_τ's, $\langle N_t^2 \rangle$ becomes proportional to t^2.) This means that the mere presence of diffusion for a periodic map indicates chaotic motion which destroys correlations within one box. We will generalize and use this characterization of chaos to some extent in Chapter 8, where we discuss area-preserving maps.

Let us finally derive a simple scaling law for the diffusion coefficient that has a purely geometric origin. If the intervals δ, through which the trajectories can move from cell to cell, are small enough (such that one can neglect the variation of ρ in this region, i.e., $\rho(x \in \delta) = \bar{\rho}$), then eq. (2.56) can be written as

$$D \approx \frac{1}{2} \bar{\rho} \delta \tag{2.57}$$

because Δ^2 has only the values zero or unity. Fig. 21 shows that D scales like

$$D \propto (a - 1)^{1/z} \tag{2.58}$$

if the map $f(x)$ has a maximum (and minimum) of order z.

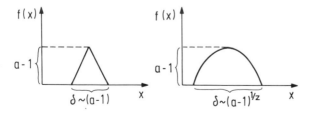

Fig. 21: The variation of δ as $(a - 1)^{1/z}$ if $f(x)$ has a maximum of order z (schematically).

3 Universal Behavior of Quadratic Maps

In this chapter, we study the logistic map

$$x_{n+1} = f_r(x_n) \equiv rx_n(1 - x_n),$$ (3.1)

shown in Fig. 22.

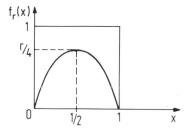

Fig. 22: The quadratic map $f_r(x)$ on the unit interval.

It has already been shown in Chapter 1 that (3.1) describes the angles x_n of a strongly damped kicked rotator. But the logistic map, which is, arguably, the simplest nonlinear difference equation, appears in many contexts.

It has already been introduced in 1845 by P.F. Verhulst to simulate the growth of a population in a closed area. The number of species x_{n+1} in the year $n + 1$ is proportional to the number in the previous year x_n and to the remaining area, which is diminished, proportionally, to x_n i.e. $x_{n+1} = rx_n(1 - x_n)$ where the parameter r depends on the fertility, the actual area of living etc.

Another example is a savings account with a self-limiting rate of interest (Peitgen and Richter, 1984). Consider a deposit z_0 which grows with a rate of interest ε as $z_{n+1} = (1 + \varepsilon)z_n = \ldots (1 + \varepsilon)^{n+1}z_0$. To prohibit unlimited wealth, some politician could suggest that the rate of interest should be reduced proportionally to z_n, i.e. $\varepsilon \rightarrow \varepsilon_0(1 - z_n/z_{max})$. Then the account develops according to $z_{n+1} = [1 + \varepsilon_0(1 - z_n/z_{max})]z_n$ which becomes equal to eq. (3.1) for $x_n = z_n\varepsilon_0/z_{max}(1 + \varepsilon_0)$ and $r = 1 + \varepsilon_0$.

One could expect for both examples that due to the feedback mechanism the quantities of interest (population and bank account) develop towards mean values. But as found by Grossmann and Thomae (1977), by Feigenbaum (1978), and by Coullet and Tresser (1978), and many others (see May, 1976, for earlier references) the iterates x_1, $x_2 \ldots$ of (3.1) display, as a function of the external parameter r, a rather complicated behavior that becomes chaotic at large r's (see Fig. 23).

Once can, therefore, understand the conclusion that May (1976) draws at the end of his article in "Nature": "Perhaps we would all be better off, not only in research and teaching, but also in everyday political and economical life, if more people would take into consideration that simple dynamical systems do not necessarily lead to simple dynamical behavior."

However, chaotic behavior is not tied to the special form of the logistic map. Feigenbaum has shown that the route to chaos that is found in the logistic map, the "Feigenbaum route", occurs (with certain restrictions which will be discussed below) in all first-order difference equations $x_{n+1} = f(x_n)$ in which $f(x_n)$ has (after a proper rescaling of x_n) only a single maximum in the unit interval $0 \le x_n \le 1$. It was found by Feigenbaum that the scaling behavior at the transition to chaos is governed by universal constants, the Feigenbaum constants α and δ, whose value depends only on the order of the maximum (e.g. quadratic, i.e. $f'(x_{max}) = 0$, $f''(x_{max}) < 0$, etc.). Because the conditions for the appearance of the Feigenbaum route are rather weak (it is practically sufficient that the Poincaré map of a system is approximately one-dimensional and has a single maximum), this route has been observed experimentally in many nonlinear systems.

The following sections of this chapter contain a rather detailed derivation of the universal properties of this route. We begin with a summary, which is intended to be a guide through the more mathematical parts.

Section 3.1 gives an overview of the numerical results for the iterates of the logistic map. It shows that the number of fixed points of $f_r(x)$ (towards which the iterates converge) doubles at distinct, increasing values of the parameter r_n. At $r = r_\infty$, the number of fixed points becomes infinite; and beyond this (finite) r-value, the behavior of the iterates is chaotic for most r's.

In Section 3.2, we investigate the pitchfork bifurcation, which provides the mechanism for the successive doubling of fixed points. It is shown that the doubling can be understood by examining the image of even iterates ($f[f(x)]$, $f[f[f[f(x)]]]$, $\ldots$) of the original map $f(x)$. This relates the generation of new fixed points to a law of functional composition. We, therefore, introduce the doubling transformation T that describes functional composition together with simultaneous rescaling along the x- and y-axis ($Tf(x) \equiv -\alpha f[f(-x/\alpha)]$) and show that the Feigenbaum constant α (which is related to the scaling of the distance between iterates) can be calculated from the (functional) fixed point f^* of T ($Tf^* = f^*$). This establishes the universal character of α. The other Feigenbaum constant δ (which measures the scaling behavior of the r_n-values) then appears as an eigenvalue of the linearized doubling transformation.

After having provided a method of calculating universal properties of the iterates, we consider several applications in Section 3.3. As a first step, we determine the relative separations of the iterates and show that the iterates form (at the accumulation point r_∞) a self-similar point set with a fractal dimensionality. We then Fourier-transform the distribution of iterates to obtain the experimentally measurable, and therefore important, power spectrum.

In any real dissipative nonlinear system, there are, due to the coupling to other degrees of freedom, also fluctuating forces, which when they are incorporated explicitly into the difference equations, tend to wash out the fine structure of the distribution of iterates. We determine the influence of this effect on the power spectrum and show that the rate at which higher subharmonics become suppressed scales via a power law with the noise level.

Up to this point, we have only considered the behavior of the iterates near the transition to chaos. It will be shown next that in the chaotic region ($r_\infty \leq r \leq 4$) periodic and chaotic r values are densely interwoven and one finds a sensitive dependence on parameter values. We also discuss the concept of structural universality and calculate the invariant density of the logistic map at $r = 4$.

Finally, in Section 3.5 we present a summary that explains the parallels between the Feigenbaum route to chaos and ordinary equilibrium second-order phase transitions. This chapter ends with a discussion of the measurable properties of the Feigenbaum route and a review of some experiments in which this route has been observed.

3.1 Parameter Dependence of the Iterates

To provide an overview in this section, we present several results for the logistic map obtained by computer iteration of eq. (3.1) for different values of the parameter r. Fig. 23 shows the accumulation points of the iterates $\{f_r^n(x_0)\}$ for $n > 300$ as a function of r together with the Liapunov exponent λ obtained via eq. (2.9).

We distinguish between a "bifurcation regime" for $1 < r < r_\infty$, where the Liapunov exponent is always negative (it becomes only zero at the bifurcation points r_n) and a "chaotic region" for $r_\infty < r \leq 4$, where λ is mostly positive, indicating chaotic behavior. The "chaotic regime" is interrupted by r-windows with $\lambda < 0$ where the sequence $\{f_r^n(x_0)\}$ is again periodic.

The numerical results can be summarized as follows:

1. Periodic regime

a) The values r_n, where the number of fixed points changes from 2^{n-1} to 2^n, scale like

$$r_n = r_\infty - \text{const.} \, \delta^{-n} \quad \text{for} \quad n \gg 1 . \tag{3.2}$$

b) The distances d_n of the point in a 2^n-cycle that are closest to $x = 1/2$ (see Fig. 24) have constant ratios:

$$\frac{d_n}{d_{n+1}} = -\alpha \quad \text{for} \quad n \gg 1 . \tag{3.3}$$

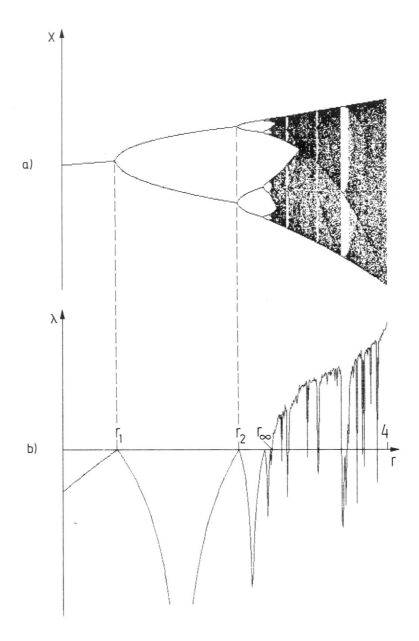

Fig. 23: a) Iterates of the logistic map, b) Liapunov exponent λ (after W. Desnizza, priv. comm.).

c) The Feigenbaum constants δ and α have the values

$$\delta = 4.6692016091... \tag{3.4a}$$

$$\alpha = 2.5029078750... \tag{3.4b}$$

Let us also note for later use that the R_n of Fig. 24 scale similar to r_n:

$$R_n - r_\infty = \text{const.}' \delta^{-n}, \tag{3.5}$$

furthermore

$$R_\infty = r_\infty = 3.5699456...$$

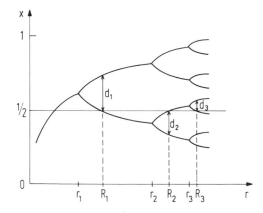

Fig. 24: Distances d_n of the fixed points closest to $x = 1/2$ for superstable 2^n-cycles (schematically).

2. Chaotic regime

a) The chaotic intervals move together by inverse bifurcations until the iterates become distributed over the whole interval $[0, 1]$ at $r = 4$.

b) The r-windows are characterized by periodic p-cycles ($p = 3, 5, 6 ...$) with successive bifurcations p, $p \cdot 2^1$, $p \cdot 2^2$ etc. The corresponding r-values scale like (3.2) with the same δ but different constants.

c) Also, period triplings $p \cdot 3^n$ and quadruplings $p \cdot 4^n$, etc. occur at $\bar{r} = \bar{r}_\infty - \overline{\text{const.}} \, \bar{\delta}^{-n}$ with different Feigenbaum constants $\bar{\delta}$, which are again universal (e.g. $\bar{\delta} = 55.247 ...$ for $p \cdot 3^n$).

3.2 Pitchfork Bifurcation and the Doubling Transformation

In this section, we show that the "Feigenbaum route" in Fig. 23 is generated by pitchfork bifurcations that relate the emergence of new branches to a universal law of functional composition. By introducing the doubling transformation T (which describes this law), we show that the Feigenbaum constants α and δ are indeed universal. They appear as the (negative inverse) value of the eigenfunction of T at $x = 1$ and as the only relevant eigenvalue of the linearized doubling operator, respectively.

Pitchfork Bifurcations

As a first step, we investigate the stability of the fixed points of $f_r(x)$ and $f_r^2(x) = f_r[f_r(x)]$ as a function of r. Fig. 25 shows that $f_r(x)$ has, for $r < 1$, only one stable fixed point at zero, which becomes unstable for $1 < r < 3$ in favor of $x^* = 1 - 1/r$.

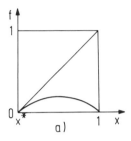
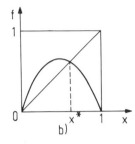

Fig. 25: The fixed points of f_r for a) $r < 1$ and b) $1 < r < 3$.

For $r > 3 = r_1$ we have $|f_r'(x^*)| = |2 - r| > 1$; i.e., x^* also becomes unstable according to criterion (2.17). What happens then?

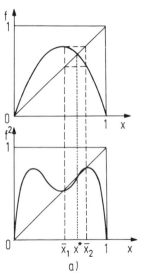
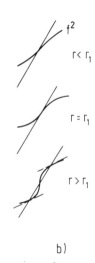

Fig. 26: a) $f(x)$ and $f^2(x) = f[f(x)]$ for $r > r_1$. b) Generation of two new stable fixed points in f^2 via a pitchfork bifurcation. (The bifurcation diagram looks like a pitchfork, see p. 182.)

Fig. 26 shows $f_r(x)$ together with its second iterate $f_r^2(x)$ for $r > r_1$. We note four properties of f^2 (the index r is dropped for convenience):

a) It has three extrema with $f^{2'} = f'[fx)]f'(x) = 0$ at $x_0 = 1/2$, because $f'(1/2) = 0$, and at $x_{1,2} = f^{-1}(1/2)$, because $f'[f[f^{-1}(1/2)]] = f'(1/2) = 0$.

b) A fixed point x^* of $f(x)$ is also a fixed point of $f^2(x)$ (and all higher iterates).

c) If a fixed point x^* becomes unstable with respect to $f(x)$, it becomes also unstable with respect to f^2 (and all higher iterates) because $|f'(x^*)| > 1$ implies $|f^{2'}(x^*)| = |f'[f(x^*)]f'(x^*)| = |f'(x^*)|^2 > 1$.

d) For $r > 3$, the old fixed point x^* in f^2 becomes unstable, and two new stable fixed points $\bar{x}_1, \bar{x}_2$ are created by a pitchfork bifurcation (see Fig. 26b).

The pair $\bar{x}_1, \bar{x}_2$ of stable fixed points of f^2 is called an attractor of $f(x)$ of period two because any sequence of iterates which starts in [0, 1] becomes attracted by $\bar{x}_1, \bar{x}_2$ in an oscillating fashion as shown in Fig. 27.

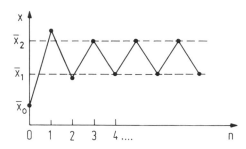

Fig. 27: Iterates of $\bar{x}_0$ if $f(x)$ has an attractor of period two (schematically).

It is easy to see that $f(x)$ maps these new fixed points of f^2 onto each other, i.e.,

$$f(\bar{x}_1) = \bar{x}_2 \text{ and } f(\bar{x}_2) = \bar{x}_1 \tag{3.6}$$

because $f^2(\bar{x}_1) = \bar{x}_1$ implies

$$ff[f(\bar{x}_1)] = f[f^2((\bar{x}_1)] = f(\bar{x}_1) \tag{3.7}$$

i.e. $f(\bar{x}_1)$ is also a fixed point of f^2, and $\bar{x}_2$ is the only possible choice. ($f(\bar{x}_1) = 0$ or x^* are at variance with $ff(\bar{x}_1) = \bar{x}_1$.)

If we now increase r beyond a value r_2, the fixed points of f^2 also become unstable. Because the derivative is the same at $\bar{x}_1$ and $\bar{x}_2$

$$f^{2'}(\bar{x}_1) = f'[f(\bar{x}_1)]f'(\bar{x}_1) = f'(\bar{x}_2)f'(\bar{x}_1) = f^{2'}(\bar{x}_2) \tag{3.8}$$

they even become unstable simultaneously.

Fig. 28 shows that after this instability the fourth iterate $f^4 = f^2 \cdot f^2$ displays two more pitchfork bifurcations which lead to an attractor of period four; i.e., one observes *period doubling*. These two examples can be generalized as follows:

a) For $r_{n-1} < r < r_n$, there exists a stable 2^{n-1}-cycle with elements x_0^*, x_1^* ... $x_{2^{n-1}-1}^*$ that is characterized by

$$f_r(x_i^*) = x_{i+1}^*, \quad f_r^{2^{n-1}}(x_i^*) = x_i^*, \quad \left| \frac{d}{dx_0^*} f_r^{2^{n-1}}(x_0^*) \right| = \left| \prod_i f_r'(x_i^*) \right| < 1 \tag{3.10}$$

b) At r_n, all points of the 2^{n-1}-cycle become unstable simultaneoulsy via pitchfork bifurcations in

$$f_r^{2^n} = f_r^{2^{n-1}} \cdot f_r^{2^{n-1}} \tag{3.11}$$

that, for $r_n < r < r_{n+1}$, lead to a new stable 2^n-cycle.

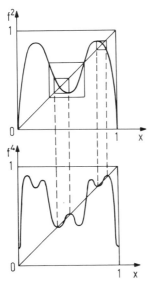

Fig. 28: Two pitchfork bifurcations in f^4 lead to an attractor of period 4.

Our last conclusion represents a first step towards universality because it connects the mechanism of subsequent bifurcations to a general law of functional composition.

Let us add as a caveat that not all quadratic maps of the unit interval onto itself display an infinite sequence of pitchfork bifurcations, but only those which have a negative Schwarzian derivative (see Appendix C).

Supercycles

To progress further, we now consider the so-called supercycles. A 2^n-supercycle is simply a superstable 2^n-cycle defined by

$$\frac{d}{dx_0^*} f_{R_n}^{2^n}(x_0^*) = \prod_i f_{R_n}'(x_i^*) = 0 \tag{3.12}$$

which implies that it always contains $x_0^* = 1/2$ as a cycle element because this is the only point where $f_r' = 0$. Referring to Fig. 24, we can see that the distances d_n are just the distances between the cycle elements $x^* = 1/2$ and $x_1 = f_{R_n}^{2^{n-1}}(1/2)$, i.e.,

$$d_n = f_{R_n}^{2^{n-1}}\left(\frac{1}{2}\right) - \frac{1}{2}.\tag{3.13}$$

In the following it is convenient to perform a coordinate transformation that displaces $x = 1/2$ to $x = 0$ such that (3.13) becomes

$$d_n = f_{R_n}^{2^{n-1}}(0).\tag{3.14}$$

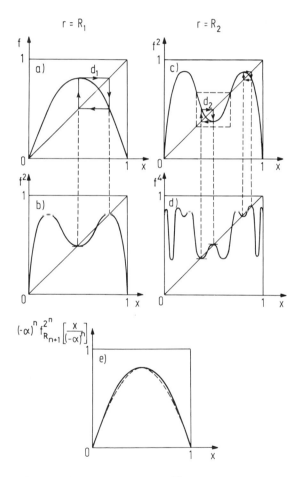

Fig. 29: The rescaled iterates $f_{R_{n+1}}^{2^n}(x)$ converge towards a universal function. a)–d) Superstable cycles at R_1 and R_2. Note the horizontal tangents in b) and d). e) The content of the dashed square of c) is rescaled (dashed line) and compared to the whole of a) (full line).

From the previous section, we see that eq. (3.3) implies

$$\lim_{n \to \infty} (-\alpha)^n d_{n+1} = d_1 \tag{3.15}$$

i.e. the sequence of scaled iterates $f_{R_{n+1}}^{2^n}(0)$ converges:

$$\lim_{n \to \infty} (-\alpha)^n f_{R_{n+1}}^{2^n}(0) = d_1 . \tag{3.16}$$

Fig. 29 suggests that (3.16) can be generalized to the whole interval, and the rescaled functions $(-\alpha)^n f_{R_{n+1}}^{2^n}[x/(-\alpha)^n]$ converge to a limiting function $g_1(x)$:

$$\lim_{n \to \infty} (-\alpha)^n f_{R_{n+1}}^{2^n}\left[\frac{x}{(-\alpha)^n}\right] = g_1(x) \tag{3.17}$$

Eq. (3.17) shows that $g_1(x)$ is determined only by the behavior of $f_{R_{n+1}}^{2^n}$ around $x = 0$ (see also Fig. 28) and should, therefore, be universal for all functions f with a quadratic maximum.

Doubling Transformation and α

As the next step, we introduce, by analogy to eq. (3.17), a whole family of functions

$$g_i(x) \equiv \lim_{n \to \infty} (-\alpha)^n f_{R_{n+i}}^{2^n}\left[\frac{x}{(-\alpha)^n}\right] ; \quad i = 0, 1 \ldots \tag{3.18}$$

We notice that all these functions are related by the *doubling transformation* T:

$$g_{i-1}(x) = (-\alpha) g_i\left[g_i\left(-\frac{x}{\alpha}\right)\right] \equiv T g_i(x) \tag{3.19}$$

because

$$g_{i-1}(x) = \lim_{n \to \infty} (-\alpha)^n f_{R_{n+i-1}}^{2^n}\left[\frac{x}{(-\alpha)^n}\right]$$

$$= \lim_{n \to \infty} (-\alpha)(-\alpha)^{n-1} f_{R_{n-1+i}}^{2^{n-1+1}}\left[-\frac{1}{\alpha}\frac{x}{(-\alpha)^{n-1}}\right]$$

$$= \lim_{m \to \infty} (-\alpha)(-\alpha)^m f_{R_{m+i}}^{2^m}\left\{\frac{1}{(-\alpha)^m}(-\alpha)^m f_{R_{m+i}}^{2^m}\left[-\frac{1}{\alpha}\frac{x}{(-\alpha)^m}\right]\right\}$$

$$= -\alpha g_i\left[g_i\left(-\frac{x}{\alpha}\right)\right] . \tag{3.20}$$

By taking the limit i → ∞ in (3.19), the function

$$g(x) \equiv \lim_{i \to \infty} g_i(x) \tag{3.21}$$

becomes a fixed point of the doubling operator T:

$$g(x) = Tg(x) = -\alpha g\left[g\left(-\frac{x}{\alpha}\right)\right]. \tag{3.22}$$

This equation determines α universally by

$$g(0) = -\alpha g[g(0)]. \tag{3.23}$$

It can easily be shown that $\mu g(x/\mu)$ is also a solution of the fixed-point equation (3.22) with the same α. Thus, the theory has nothing to say about absolute scales, and we fix μ by setting

$$g(0) = 1. \tag{3.24}$$

Although a general theory for the solution of the functional equation (3.22) is still lacking, we can obtain a unique solution if we specify the nature of the maximum of $g(x)$ at $x = 0$ (for example quadratic) and require that $g(x)$ is a smooth function. If we use for $g(x)$ in the quadratic case the extremely short power law expansion

$$g(x) = 1 + bx^2 \tag{3.25}$$

the fixed point equation (3.22) becomes

$$1 + bx^2 = -\alpha(1 + b) - \left(\frac{2b^2}{\alpha}\right)x^2 + O(x^4) \tag{3.26}$$

which yields

$$b = (-2 - \sqrt{12})/4 \simeq -1.366; \quad \alpha = |2b| \simeq 2.73. \tag{3.27}$$

These values only differ by 10% from Feigenbaum's numerical results

$$g(x) = 1 - 1.52763 x^2 + 0.104815 x^4 + 0.0267057 x^6 - \cdots$$

$$\alpha = 2.502807876 \ldots \tag{3.28}$$

This establishes the universality of α.

Linearized Doubling Transformation and δ

What can we say about the scaling along the r-axis? The values $r = R_n$, for which a 2^n-cycle becomes superstable, are determined by the condition that $x = 1/2$ is an element of the supercycle (see eq. (3.12), i.e., $x = 1/2$ is a fixed point of $f_{R_n}^{2^n}(x)$:

$$f_{R_n}^{2^n}\left(\frac{1}{2}\right) = \frac{1}{2} \tag{3.29}$$

which after translation by $1/2$ becomes (see eqns. (3.13–14)):

$$f_{R_n}^{2^n}(0) = 0 . \tag{3.30}$$

This equation has a large number of solutions because it also yields the 2^n-supercycles that occur in the windows of the chaotic regime. In order to single out the R_n-values in the bifurcation region with

$$r_1 < R_1 < r_2 < R_2 < r_3 \ldots , \tag{3.31}$$

(3.30) is solved starting from $n = 0$, and the R_n are ordered as in (3.31).

The R_n tell us how R_∞ is approached. In order to prove the scaling relation (3.5),

$$R_n - R_\infty \propto \delta^{-n} , \tag{3.32}$$

we expand $f_R(x)$ around $f_{R_\infty}(x)$:

$$f_R(x) = f_{R_\infty}(x) + (R - R_\infty)\delta f(x) + \ldots$$

where

$$\delta f(x) \equiv \left. \frac{\partial f_R(x)}{\partial R} \right|_{R_\infty} . \tag{3.33}$$

Let us now apply the doubling operator T to this equation. A straightforward linearization in δf yields

$$\mathrm{T}f_R = \mathrm{T}f_{R_\infty} + (R - R_\infty)\mathrm{L}_{f_{R_\infty}}\delta f + \mathrm{O}[(\delta f)^2] \tag{3.34}$$

where L_f is the linear operator

$$\mathrm{L}_f \delta f = -\alpha \left\{ f'\left[f\left(-\frac{x}{\alpha}\right)\right]\delta f\left(-\frac{x}{\alpha}\right) + \delta f\left[f\left(-\frac{x}{\alpha}\right)\right] \right\} . \tag{3.35}$$

Note that L_f is only defined with respect to a function f.

Repeated application of T yields

$$T^n f_R = T^n f_{R_\infty} + (R - R_\infty) L_{T^{n-1} f_{R_\infty}} \ldots L_{f_{R_\infty}} \delta f + O\left[(\delta f)^2\right] . \tag{3.36}$$

We observe that, according to eqns. (3.18–21), $T^n f_{R_\infty}$ converges to the fixed point,

$$T^n f_{R_\infty}(x) = (-\alpha)^n f_{R_\infty}^{2^n} \left[\frac{x}{(-\alpha)^n} \right] \cong g(x) \quad \text{for} \quad n \gg 1 , \tag{3.37}$$

and (3.36) becomes approximately:

$$T^n f_R(x) \cong g(x) + (R - R_\infty) L_g^n \delta f(x) \quad \text{for} \quad n \gg 1 . \tag{3.38}$$

This equation can be further simplified if we expand $\delta f(x)$ with respect to the eigenfunctions φ_ν of L_g,

$$L_g \varphi_\nu = \lambda_\nu \varphi_\nu ; \quad \delta f = \sum_\nu c_\nu \varphi_\nu ; \quad \nu = 1, 2 \ldots \tag{3.39}$$

$$\rightarrow L_g^n \delta f = \sum_\nu c_\nu \lambda_\nu^n \varphi_\nu \tag{3.40}$$

and assume that only one of the eigenvalues λ_ν is larger than unity, i.e.,

$$\lambda_1 > 1 ; \quad |\lambda_\nu| < 1 \quad \text{for} \quad \nu \neq 1 . \tag{3.41}$$

We then obtain only the contribution from λ_1 in (3.40),

$$L_g^n \delta f \cong c_1 \lambda_1^n \varphi_1 \quad \text{for} \quad n \gg 1 , \tag{3.42}$$

and (3.38) reduces to

$$T^n f_{R_n}(x) \cong g(x) + (R - R_\infty) \cdot \delta^n \cdot a \cdot h(x) \quad \text{for} \quad n \gg 1 \tag{3.43}$$

where we introduced $c_1 \equiv a$, $\varphi_1 \equiv h$, $\lambda_1 \equiv \delta$.

The eigenvalue $\lambda_1 \equiv \delta$ is identical with Feigenbaum's constant because for $R = R_n$ and $x = 0$, (3.43) yields

$$T^n f_{R_n}(0) = g(0) + (R_n - R_\infty) \cdot \delta^n \cdot a \cdot h(0) \tag{3.44}$$

and from (3.30) we have the condition

$$T^n f_{R_n}(0) = (-\alpha)^n f_{R_n}^{2^n}(0) = 0 . \tag{3.45}$$

This leads to the desired result (note $g(0) = 1$)

$$\lim_{n \to \infty} (R_n - R_\infty) \cdot \delta^n = \frac{-1}{a \cdot h(0)} = \text{const} . \tag{3.46}$$

The last equation can be generalized if we introduce the slopes

$$\mu \equiv \frac{d}{dx_0^*} f_r^{2^n}(x_0^*) = \prod_i f_r'(x_i^*)$$

(3.47)

as a parameter and characterize r by the pair (n, μ), as shown in Fig. 30.

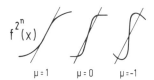

Fig. 30: Parametrization of r by n and μ (schematically), i. e. $r_n = R_{n,1} \triangleq (n, 1)$ and $R_n = R_{n,0} \triangleq (n, 0)$.

Then we obtain from (3.44):

$$\lim_{n \to \infty} (R_{n,\mu} - R_\infty) \cdot \delta^n = \frac{g_{0,\mu}(0) - g(0)}{\alpha \cdot h(0)}$$

(3.48a)

where

$$g_{0,\mu}(x) = \lim_{n \to \infty} (-\alpha)^n f_{R_{n,\mu}}^{2^n} \left[\frac{x}{(-\alpha)^n} \right]$$

(3.48b)

is again a universal function of μ.

At the bifurcation points, r_n, the slopes have always the same value $\mu = 1$ (see Fig. 30). Therefore, the r_n's scale according to (3.48) with the same δ as the R_n's of the superstable cycles (with $\mu = 0$):

$$r_n - r_\infty \propto \delta^{-n} \quad \text{for} \quad n \gg 1 \,.$$

(3.49a)

Note that the accumulation point is the same for all μ's:

$$\lim_{n \to \infty} R_{n,\mu} = R_\infty = r_\infty$$

(3.49b)

because $r_n \leq R_{n,\mu} \leq r_{n+1}$ and $r_{n+1} - r_n \to 0$ for $n \to \infty$.

The numerical value for δ can be obtained (by combining (3.35–43)) from the universal eigenvalue equation

$$L_g h(x) = -\alpha \left\{ g' \left[g \left(-\frac{x}{\alpha} \right) \right] h \left(-\frac{x}{\alpha} \right) + h \left[g \left(-\frac{x}{\alpha} \right) \right] \right\} = \delta \cdot h(x) \,.$$

(3.50)

To make things simple we retain in the power law expansion for $h(x)$ only the first term $h(0)$ such that (3.50) becomes an algebraic equation for δ:

$$-\alpha \{g'[g(0)] + 1\} = \delta . \tag{3.51a}$$

The value $g'[g(0)] = g'(1)$ follows for functions with a quadratic maximum (i.e. $g''(0) \neq 0$) by differentiating the fixed-point equation (3.22) twice:

$$g''(x) = -\left\{g''\left[g\left(-\frac{x}{\alpha}\right)\right]\left[g'\left(-\frac{x}{\alpha}\right)\right]^2 + g'\left[g\left(-\frac{x}{\alpha}\right)\right]g''\left(-\frac{x}{\alpha}\right)\right\}/\alpha$$

$$\to g'(1) = -\alpha . \tag{3.51b}$$

Thus (3.51 a) becomes

$$\delta = \alpha^2 - \alpha . \tag{3.51c}$$

(For functions with a maximum of order $2z$ one finds $\delta = \alpha^{1+z} - \alpha$.)

Using our previously determined value $\alpha = 2.73$, we obtain $\delta \approx 4.72$ from (3.51), i.e., an accuracy of about 1% with respect to Feigenbaum's numerical result $\delta = 4.6692016\ldots$ This is not so bad if one considers the crudeness of our approximation.

It is of course much more laborious to show that δ is indeed the only eigenvalue of L_g which is larger than unity. Extensive computer calculations by Feigenbaum and the analytical results of Collet, Eckmann, and Lanford (1980) have proven this assumption.

Summarizing, the two main results of this section are

a) the fixed-point equation for the doubling operator (3.22)

$$Tg(x) = -\alpha g\left[g\left(-\frac{x}{\alpha}\right)\right] = g(x) \tag{3.52}$$

which establishes the universality of α,

b) the linearized doubling transformation (3.43)

$$T^n f_R(x) = g(x) + (R - R_\infty) \cdot \delta^n \cdot a \cdot h(x) \quad \text{for} \quad n \gg 1 \tag{3.53}$$

which shows that δ is universal and determines the way in which a function is repelled from the fixed-point function $g(x)$.

Universality emerges here because the linearized doubling operator L_g has only one *relevant* eigenvalue $\lambda_1 > 1$ such that all functions $f(x)$ — with the exception of $\varphi_1(x)$ — renormalize, after several applications of T, to the fixed-point function $g(x)$ because the eigenvalues belonging to $f - g = \sum_{\nu \neq 1} c_\nu \varphi_\nu$ are smaler than unity, i.e., *irrelevant*.

3.3 Self-Similarity, Universal Power Spectrum, and the Influence of External Noise

In this section, we calculate the distances between the elements of a 2^n-cycle and determine its power spectrum. It is then shown that external noise changes the power spectrum drastically and destroys higher subharmonics. Finally, we discuss the bifurcation diagram for $r > r_\infty$ and show that the chaotic behavior of the iterates (of the logistic map) at $r = 4$ is related to the chaos of a triangular map.

The power spectrum is an important tool for characterizing irregular motion. In order to calculate this quantity for a system that exhibits the Feigenbaum route to chaos, we identify the time variable with n and determine as a first step the relative positions of the cycle elements.

Self-Similarity in the Positions of the Cycle Elements

All we know up to now about the positions of the cycle elements is that according to eqns. (3.3) and (3.14) the distances $d_n(0)$ of the supercycle elements closest to $x = 0$ scale with α, i.e.

$$\frac{d_{n+1}(0)}{d_n(0)} = -\frac{1}{\alpha} \quad \text{for} \quad d_n(0) = f_{\bar{R}_n}^{2^{n-1}}(0), n \gg 1 \,. \tag{3.54}$$

It is now our aim to generalize these equations. We will calculate for all m the distance $d_n(m)$ of the mth element x_m of a 2^n-supercycle to its nearest neighbor $f_{\bar{R}_n}^{2^{n-1}}(x_m)$,

$$d_n(m) \equiv x_m - f_{\bar{R}_n}^{2^{n-1}}(x_m) \tag{3.55}$$

and the change of $d_n(m)$ if one increases n,

$$\sigma_n(m) \equiv \frac{d_{n+1}(m)}{d_n(m)} \,. \tag{3.56}$$

The function $\sigma_n(m)$ changes sign after 2^n cycle steps,

$$\sigma_n(m + 2^n) = -\sigma_n(m) \tag{3.57}$$

because

$$d_{n+1}(m + 2^n) = f_{\bar{R}_{n+1}}^{2^n}(x_m) - f_{\bar{R}_{n+1}}^{2^n}[f_{\bar{R}_{n+1}}^{2^n}(x_m)]$$
$$= f_{\bar{R}_{n+1}}^{2^n}(x_m) - x_m = -d_{n+1}(m) \tag{3.58}$$

and $d_n(m)$ is left invariant ($f_{\bar{R}_n}^{2^n}(x_m) = x_m$).

Let us now consider the values $m = 2^{n-i}$, $i = 0 \ldots n$, and evaluate $\sigma_n(m)$ in the limit $n \gg 1$. The definitions (3.55), (3.56) yield

$$\sigma_n[2^{n-i}] = \frac{f_{\tilde{R}_{n+1}}^{2^{n-i}}(0) - f_{\tilde{R}_{n+1}}^{2^n}[f_{\tilde{R}_{n+1}}^{2^{n-i}}(0)]}{f_{\tilde{R}_n}^{2^{n-i}}(0) - f_{\tilde{R}_n}^{2^{n-1}}[f_{\tilde{R}_n}^{2^{n-i}}(0)]} \tag{3.59}$$

$$= \frac{f_{\tilde{R}_{(n-i)+i+1}}^{2^{n-i}}(0) - f_{\tilde{R}_{(n-i)+i+1}}^{2^{n-i}}[f_{\tilde{R}_{n+1}}^{2^n}(0)]}{f_{\tilde{R}_{(n-i)+i}}^{2^{n-i}}(0) - f_{\tilde{R}_{(n-i)+i}}^{2^{n-i}}[f_{\tilde{R}_{(n-1)+1}}^{2^{n-1}}(0)]}$$

and because

$$f_{\tilde{R}_{l+j}}^{2^l}(x) \cong (-\alpha)^{-l} g_j[(-\alpha)^l x] \quad \text{for} \quad l = n - i \to \infty \tag{3.60}$$

this becomes

$$\sigma_n[2^{n-i}] = \frac{g_{i+1}(0) - g_{i+1}[(-\alpha)^{-i} g_1(0)]}{g_i(0) - g_i[(-\alpha)^{-i+1} g_1(0)]} \quad \text{for} \quad n \gg 1. \tag{3.61}$$

We note that the functions $g_i(x)$ can be obtained from (3.18) and (3.44) for $i \gg 1$:

$$g_i(x) = \lim_{n \to \infty} T^n f_{R_{n+i}}(x) = g(x) - \delta^{-i} \cdot h(x). \tag{3.62}$$

For smaller i one uses the recursion (3.19),

$$g_{i-1}(x) = T g_i(x). \tag{3.63}$$

If we introduce, for convenience, the new variable $x = \dfrac{m}{2^{n+1}}$ and drop the index n, the symmetry relation (3.57) reads

$$\sigma\left(x + \frac{1}{2}\right) = -\sigma(x). \tag{3.64}$$

This generates from our familiar scaling relation (3.54) the value of σ at $x = 1/2$:

$$\sigma(0) = \frac{-1}{\alpha} \to \sigma\left(\frac{1}{2}\right) = -\sigma(0) = \frac{1}{\alpha}. \tag{3.65}$$

But starting instead from (3.61) we obtain

$$\sigma(0^+) = \lim_{i \to \infty} \sigma(x = 2^{-1-i})$$

$$= \lim_{i \to \infty} \frac{g(0) - g[(-\alpha)^{-i} g_1(0)]}{g(0) - g[(-\alpha)^{-i+1} g_1(0)]} = \frac{1}{\alpha^2} \tag{3.66}$$

because

$$g[(-\alpha)^{-i} g_1(0)] \approx g(0) + \frac{1}{2} g''(0)(-\alpha)^{-2i} g_1^2(0) \tag{3.67}$$

and from there and (3.64):

$$\sigma\left(\frac{1}{2} + 0^+\right) = -\sigma(0^+) = -\frac{1}{\alpha^2}. \tag{3.68}$$

This means that $\sigma(x)$ is discontinuous at $x = 0$ and $x = 1/2$.

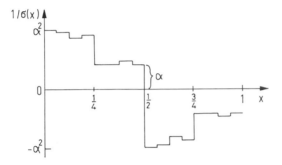

Fig. 31: The function $\sigma(x) = \sigma_n(m = 2^n \cdot x)$, $n \gg 1$ (see eq. (3.56)) tells how distances between adjacent points change if one passes, for $n \gg 1$, from a 2^n- to a 2^{n+1}-superstable cycle (after Feigenbaum, 1980).

More elaborate calculations show that $\sigma(x)$ jumps at all rationals as depicted in Fig. 31. Fortunately, the discontinuities decrease rapidly as the number of terms in the binary expansion of the rational increases, and it is therefore often sufficient to consider only the jumps at $x = 0$, $1/4$, $1/2$.

Hausdorff Dimension

According to Fig. 31, the distances between nearby points in a supercycle change with universal ratios after each bifurcation. The self-similarity of this pattern can be characterized by the Hausdorff dimension of the attractor.

If for a set of points in d dimensions the number $N(l)$ of d-spheres of diameter l needed to cover the set increases like

$$N(l) \propto l^{-D} \quad \text{for} \quad l \to 0 \tag{3.69}$$

then D is called the *Hausdorff* dimension of the set. (The quantity defined in eq. (3.69) is actually the capacity dimension which agrees for our purposes with the Hausdorff dimension whose rigorous definition is, e.g., elaborated in Falconer's book on the *Geometry of Fractal Sets* (1985)).

For the self-similar sets shown in Fig. 32, D can be calculated from

$$D = -\frac{\log [N(l)/N(l')]}{\log (l/l')} .$$

(3.70)

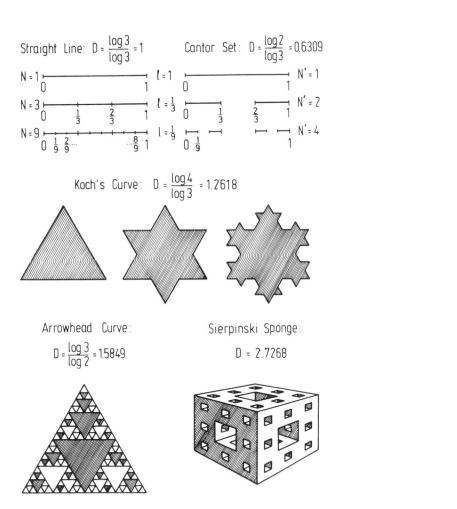

Fig. 32: Hausdorff dimension of a straight line and of some typical self-similar point sets, so-called *fractals* (drawn after Mandelbrot, 1982). It is understood that the ramifications continue ad infinitum. Koch's curve is a line of infinite length that encloses a finite area.

We note that the length L of the Cantor set shown in Fig. 32 is indeed zero:

$$L = 1 - \frac{1}{3} - \frac{2}{9} - \frac{4}{27} \cdots = 1 - \frac{1}{3} \sum_{v=0}^{\infty} \left(\frac{2}{3}\right)^v = 0 .$$ (3.71)

The Hausdorff dimension D^* of a 2^n-cycle can be calculated in the limit $n \to \infty$ as follows. If for a 2^n-supercycle we need $N(l) = 2^n$ segments of length l to cover all its points, then from Fig. 31 it is found that the mean minimum length l' to cover all $N(l') = 2^{n+1}$ cycles is given approximately by

$$l' \approx \frac{1}{2^{n+1}} \left[2^n \frac{l}{\alpha} + 2^n \frac{l}{\alpha^2} \right]$$ (3.72)

which yields

$$D^* = - \log 2/\log \left[\frac{1}{2} \left(\frac{1}{\alpha} + \frac{1}{\alpha^2} \right) \right] \cong 0.543 .$$ (3.73)

This value differs only by 5% from Grassberger's (1981) analytical and numerical result $D^* = 0.5388 \ldots$ (The numerical result was obtained by covering the attractor with successively smaller segments l and counting $N(l)$).

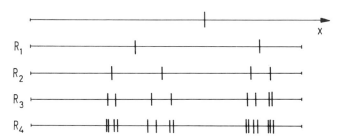

Fig. 33: Positions of the cycle elements for $f_{R_n}(x) = R_n x (1 - x)$.

Fig. 33 demonstrates the typical Cantor-set structure of the attractor. We will now show that this leads to a remarkably simple change in the measurable power spectrum after each bifurcation step.

Power Spectrum

The power spectrum $P(k)$ can be obtained by resolving the element $x^n(t) \equiv f_{R_n}^t(0)$ of a 2^n-cycle ($t = 1, 2, \ldots, 2^n \equiv T_n$) into its Fourier components a_k^n

$$x^n(t) = \sum_k a_k^n\, e^{\frac{2\pi i k}{T_n} t} .$$ (3.74)

The periodicity of the cycle implies

$$x^n(t) = x^n(t + 2^n) \rightarrow e^{2\pi i k} = 1 \rightarrow k = 0, 1, \ldots, 2^n - 1 \tag{3.75}$$

i.e. after each bifurcation step from $n \rightarrow n + 1$, 2^n new subharmonics with frequencies $k/2^{n+1}$ ($k = 1, 3, 5, \ldots$) are obtained, as shown in Fig. 34. The corresponding change in the a_k^n's can be calculated from $\sigma(m)$.

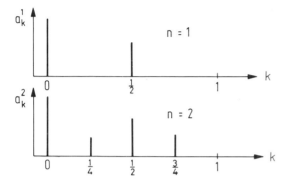

Fig. 34: Change of the Fourier components after one bifurcation (schematically).

As a first step, we invert (3.74):

$$a_k^n - \frac{1}{2^n} \sum_{t=1}^{2^n} e^{\frac{2\pi i k t}{2^n}} x^n(t) \approx \frac{1}{T_n} \int_0^{T_n} dt \, e^{\frac{2\pi i k t}{T_n}} x^n(t) \tag{3.76}$$

and by splitting the interval $[0, T_{n+1}]$ into two halves with $T_n = \frac{1}{2} T_{n+1}$, we obtain:

$$a_k^{n+1} = \int_0^{T_n} \frac{dt}{2T_n} [x^{n+1}(t) + (-1)^k x^{n+1}(t + T_n)] e^{-\frac{\pi i k t}{T_n}}. \tag{3.77}$$

The new even harmonics a_{2k}^{n+1} are essentially represented by the old spectrum at n (see Fig. 34), because

$$a_{2k}^{n+1} = \int_0^{T_n} \frac{dt}{2T_n} [x^{n+1}(t) + x^{n+1}(t + T_n)] e^{-\frac{2\pi i k t}{T_n}}$$

$$\approx \int_0^{T_n} \frac{dt}{T_n} x^{n+1}(t) e^{-\frac{2\pi i k t}{T_n}} \approx \int_0^{T_n} \frac{dt}{T_n} x^n(t) e^{-\frac{2\pi i k t}{T_n}} = a_k^n. \tag{3.78}$$

The calculation of the odd components is somewhat more delicate, and we require our previously calculated function $\sigma(x)$.

From (3.77) we have

$$a_{2k+1}^{n+1} = \int_0^{T_n} \frac{dt}{2T_n} \, [x^{n+1}(t) - x^{n+1}(t + T_n)] \, e^{-\frac{(2k+1)\pi i t}{T_n}} \tag{3.79}$$

and

$$x^{n+1}(t) - x^{n+1}(t + T_n) = x^{n+1}(t) - f_{R_{n+1}}^{2^n}[x^{n+1}(t)] = d^{n+1}(t)$$

$$= \sigma\left(\frac{t}{2T_n}\right) d^n(t) \tag{3.80}$$

with

$$d^n(t) = x^n(t) - x^n(t + T_{n-1}) = \sum_k a_k^n [1 - (-1)^k] \, e^{\frac{2\pi i k t}{T_n}}$$

$$= 2 \sum_k a_{2k+1}^n \, e^{\frac{2\pi i (2k+1)}{T_n}} \ . \tag{3.81}$$

Thus, we obtain:

$$a_{2k+1}^{n+1} = \sum_{k'} a_{2k'+1}^n \int_0^{T_n} \frac{dt}{T_n} \, \sigma\left(\frac{t}{2T_n}\right) e^{\frac{2\pi i t}{T_n}[2k'+1 - \frac{1}{2}(2k+1)]}$$

$$\approx \sum_{k'} a_{2k'+1}^n \left[\frac{1}{\alpha^2} + \frac{1}{\alpha} + i(-1)^k \left(\frac{1}{\alpha} - \frac{1}{\alpha^2} \right) \right] \frac{1}{2\pi i} \frac{1}{2k'+1 - \frac{1}{2}(2k+1)} \tag{3.82}$$

because

$$\int_0^1 d\xi \, \sigma\left(\frac{\xi}{2}\right) e^{2\pi i \xi y} \approx \frac{1}{\alpha^2} \int_0^{1/2} d\xi \, e^{2\pi i \xi} + \frac{1}{\alpha} \int_{1/2}^1 d\xi \, e^{2\pi i \xi y}$$

$$= \frac{1}{2\pi i} \frac{1}{y} \, [(e^{\pi i y} - 1)/\alpha^2 + (e^{2\pi i y} - e^{\pi i y})/\alpha] \tag{3.83}$$

where $\sigma(x)$ is approximated by a simple piecewise constant function.
Replacing the sum over k' in (3.82) by an integral and using

$$\frac{1}{2\pi i} \int dk' \, x_{2k'+1}^n \frac{1}{2k'+1 - \frac{1}{2} + (2k+1)} \approx \frac{1}{4} x_{(1/2)(2k+1)}^n \tag{3.84}$$

we eventually obtain:

$$|a_{2k+1}^{n+1}| \approx \mu^{-1} |a_{(1/2)(2k+1)}^{n}| \, , \quad \mu^{-1} = \frac{1}{4\alpha} \sqrt{2 \left(1 + \frac{1}{\alpha^2} \right)} \tag{3.85}$$

$$\mu^{-1} = 0.1525 \, , \quad \text{i.e.} \quad 10 \log_{10} \mu = 8.17 \, \text{dB} \, .$$

Therefore, the amplitudes of the odd subharmonics, which appear after each bifurcation step, are "in the mean" just the averaged amplitudes of the old odd components reduced by a constant factor μ^{-1}. (The many approximations which habe been made in deriving (3.85) require this cautious restriction to averages.) The universal pattern

$$|a_{2k}^{n+1}| \approx |a_k^n| \, , \quad |a_{2k+1}^{n+1}| \approx 0.152 \, |a_{(1/2)(2k+1)}^n| \tag{3.86}$$

is shown schematically in Fig. 34 and is reasonably consistent, e. g., with the numerical result found for the quadratic map $f(x) = 1 - 1.401155 \, x^2$ depicted in Fig. 35.

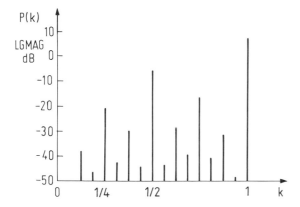

Fig. 35: Numerically determined power spectrum for a quadratic map. Subsequent odd subharmonics differ by a factor μ^{-1} (after Collet and Eckmann, 1980).

Influence of External Noise

The full details of this power spectrum cannot be observed experimentally because there will always be some external noise due to the coupling to other degrees of freedom (see Fig. 36). In order to discuss this perturbation quantitatively, we add a noise term ξ_n to the logistic equation:

$$x_{n+1} = f_r(x_n) + \xi_n \tag{3.87}$$

and calculate its influence on the cascade of bifurcations.

Here, ξ_n are Gaussian-distributed variables with averages

$$\langle \xi_n \xi_{n'} \rangle = \sigma^2 \delta_{n,n'} \tag{3.88}$$

(similarly their Fourier components ξ_k are Gaussian-distributed), and σ measures the intensity of the white noise. We recall that the new Fourier components $|a_k^{n+1}|$ of a 2^{n+1}-cycle are a factor of μ^{-1} smaller than the old components $|a_k^n|$. This means that any finite external noise eventually suppresses all subharmonics above a certain n, as shown in Fig. 36c.

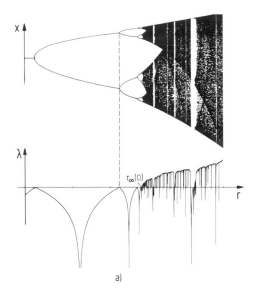

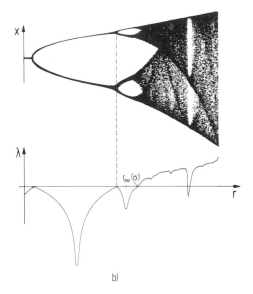

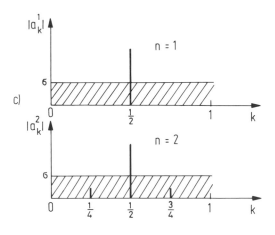

Fig. 36: a) Iterates of the logistic map and its Liapunov exponent λ, compared to b) the corresponding quantities in the presence of external noise with amplitude $\sigma = 10^{-3}$ (after Crutchfield, Farmer and Huberman, 1982). Although the noise washes out the fine structure in the iterates and in λ, there is still a sharp transition to chaos which is indicated by the change of sign of λ in b). c) Suppression of subharmonics in the presence of white noise σ. Note that the subharmonic amplitudes are reduced by a factor μ^{-1} for $n \to n + 1$ (schematically).

In fact the values R_n (above which all subharmonics become unobservable because they have merged into the chaos provided by the external noise) and the corresponding amplitude σ_n are related by a power law

$$(R_\infty - R_n) \propto \sigma_h^\gamma \tag{3.89}$$

where $\gamma = \log \delta / \log \mu$.

This can be derived as follows: If at R_1 a noise level σ_1 is just sufficient to suppress the first subharmonic $|a_k^1|$, then all $|a_k^n| = \mu^{-n} |a_k^1|$ will disappear at R_n for $\sigma_n = \mu^{-n} \sigma_1$. If the common n is eliminated, the corresponding scaling relations

$$(R_\infty - R_n) \propto \delta^{-n} \tag{3.90a}$$

$$\sigma_n \propto \mu^{-n} \tag{3.90b}$$

yield (3.89).

The decrease of R_n with increasing noise amplitude as in (3.89) has been verified numerically as shown in Fig. 37.

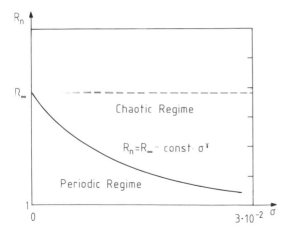

Fig. 37: Suppression of the periodic regime by the presence of external noise for the logistic map (after Crutchfield, Farmer and Huberman, 1982).

The external noise, which produces chaos for $R_n < R_\infty$, plays a similar role as a magnetic field which causes a finite magnetization above the critical point of a magnet. This analogy has been worked out by Shraiman, Wayne, and Martin (1981) who have shown, for example, that the Liapunov exponent scales in the presence of external noise like

$$\lambda = r^\beta \lambda_0 [r^{-1/\gamma} \sigma] \; ; \quad \beta = \log 2 / \log \delta \; ; \quad r = R_\infty - R \tag{3.91a}$$

or equivalently

$$\lambda = \sigma^\theta \lambda_1 [r\sigma^{-\gamma}] \; ; \quad \theta = \log 2/\log \mu \tag{3.91 b}$$

where $\lambda_{0,1}$ are universal functions (see Fig. 38). These results have also been obtained by Feigenbaum and Hasslacher (1982) using a decimation of path integrals. Their method, which has a wide range of potential applications, is explained in Appendix E. Eq. (3.91 a) is reminiscent of the scaling behavior of the magnetization M at a second-order phase transition:

$$M = r^\beta f(r^{1/\gamma} h) \tag{3.92}$$

where $r = |T - T_c|$ is the temperature distance to the critical point, and h is the magnetic field. For the onset of chaos, where λ changes sign, eq. (3.91 a) yields

$$0 = \lambda_0 [r^{-1/\gamma} \sigma] \rightarrow r^{-1/\gamma} \cdot \sigma = \text{const.} \tag{3.93}$$

i. e. our equation (3.89).

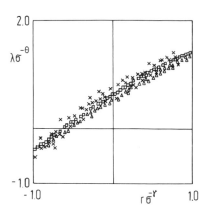

Fig. 38: Numerical determination of the scaling function $\lambda_1 (\gamma)$ in eq. (3.91 b). The quantity $\lambda \sigma^{-\theta}$ is plotted against 100 values of $y = r\sigma^{-\gamma}$ at each of three noise levels: $\sigma = 10^{-6}$ ($\square$), 10^{-8} ($\triangle$), and 10^{-10} ($\times$) (after Crutchfield et al., 1981).

3.4   Behavior of the Logistic Map for $r_\infty \leq r$

Let us now discuss the behavior of the logistic map for $r \geq r_\infty$. We have already seen above that at r_∞ the sequence of bifurcations ends in a set of infinitely many points, the so-called Feigenbaum attractor, which has a Hausdorff dimension $D = 0.548 \ldots$ Fig. 23 shows that the Liapunov exponent λ of the logistic map at r_∞ is still zero, i. e. the Feigenbaum attractor is no strange attractor (see Chapter 5 for the definition of this object). But according to Fig. 23, λ becomes mostly positive for $r > r_\infty$, and it is therefore reasonable to say that chaos starts at the end of the bifurcation region. Although the detailed behavior of the iterates (of the logistic map) appears rather complicated in this region, it shows regularities which are again dictated by the doubling operator and therefore universal. It will be shown in the first part of this section that for $r_\infty < r$, periodic and chaotic regions are densely interwoven, and one finds a sensitive dependence on the parameter values. Next we discuss the structural universality discovered by Metropolis, Stein and Stein (1973) which preceeded the work of Feigenbaum (1978). Finally we calculate the invariant density at $r = 4$ and explain the scaling of the reverse band-splitting bifurcations.

Sensitive Dependence on Parameters

Fig. 32 shows that for $r_\infty < r \leqslant 4$ "chaotic parameter values" r with $\lambda > 0$ and non-chaotic r's with $\lambda < 0$ are densely interwoven. Close to every parameter value where there is chaos, one can find another r value which corresponds to a stable periodic orbit, that is, the logistic map displays a sensitive dependence on the parameter r. The practical implications of this behavior are worse than those of sensitive dependence on initial conditions. When chaos occurs, the only alternative is to resort to statistical predictions. But for sensitive dependence on parameters, statistical averages become unstable under variations in parameters because the average behavior of the system may be completely different in the periodic and in the chaotic case.

Although there is a rigorous proof (Jacobson, 1981) that the total length of chaotic parameter intervals in $r_\infty \leqslant r \leqslant 4$ is finite, there remain the following questions:

— Which fraction of parameter values is chaotic?
— What is the probability that a change in the parameter values will lead to a change in qualitative behavior. Since it is no longer possible to distinguish experimentally (i. e. when one has finite precision) between chaotic and nonchaotic parameter values, one can only make statistical predictions for the parameter dependence of the system.

An answer to these questions has been given by D. Farmer (1985) who calculated numerically the coarse grained measure (i. e. total length) $\mu(l)$ of all chaotic parameter intervals for $f(x) = rx(1 - x)$ and $g(x) = r\sin(\pi x)$. Coarse grained means that all nonchaotic holes on the r axis, with a size larger than l, were deleted (see Fig. 39).

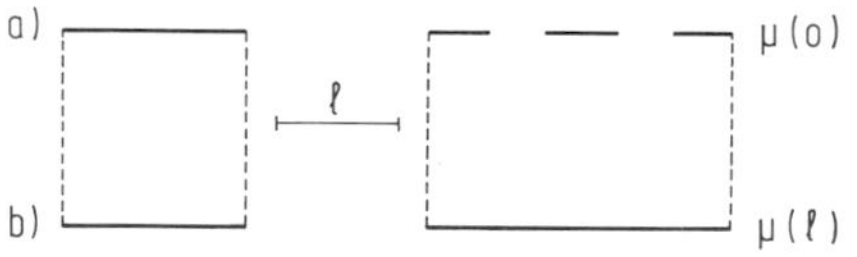

Fig. 39: a) Piece of a fat fractal with measure $\mu(0)$, b) its coarse grained measure $\mu(l)$ is larger than $\mu(0)$ because only those holes that are bigger than the resolution l are deleted.

Fig. 40 shows that $\mu(l)$ scales like

$$\mu(l) = \mu(0) + Al^\beta \tag{3.94}$$

where $\beta = 0.45 \pm 0.05$ is numerically the same for both maps, whereas $\mu(0) = 0.8979$ (0.8929) for $f(x)$ and $g(x)$ respectively.

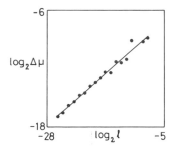

Fig. 40: The logarithm of the change $\Delta\mu(l) = \mu(l) - \mu(0)$ in the coarse grained measure of chaotic parameter intervals plotted against the logarithm of the resolution l (after Farmer, 1985).

The set of chaotic parameter values with a scaling behavior described by eq. (3.94) is an example of a "fat" fractal. Fat fractals have, in contrast to the "thin" fractals considered on page 55, a finite measure (i.e. volume). A typical example of a fat Cantor set is shown in Fig. 41 which is obtained by deleting from the unit interval the central $1/3$, $1/9$, $1/27$ ... of each piece.

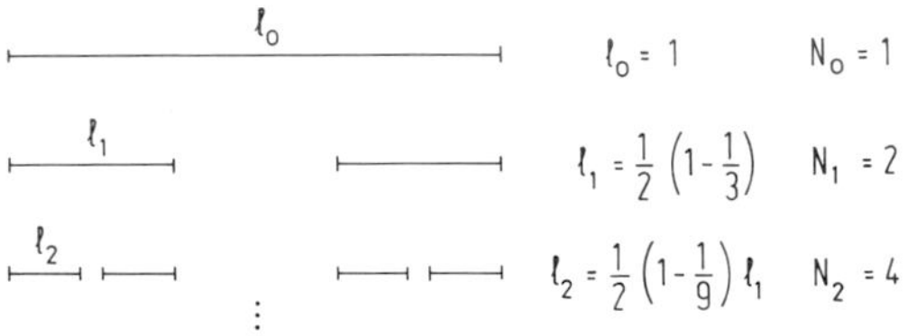

$$l_0 = 1 \qquad N_0 = 1$$

$$l_1 = \frac{1}{2}\left(1 - \frac{1}{3}\right) \qquad N_1 = 2$$

$$l_2 = \frac{1}{2}\left(1 - \frac{1}{9}\right)l_1 \qquad N_2 = 4$$

Fig. 41: Fat fractal which is constructed by deleting the central $1/3$ then $1/9$ ... of each remaining subinterval (compare this to the thin fractal in Fig. 32).

The remaining lengths scale like $l_n = 1/2\,[1 - (1/3)^n]\,l_{n-1}$. Using $N_n = 2^n$, the Hausdorff dimension D of this fat Cantor set becomes $D = 1$ via eq. (3.70). However, its volume scales according to:

$$\mu(l_n) - \mu(0) = N_n l_n - N_\infty l_\infty = \prod_{j=1}^{n}\left[1 - \left(\frac{1}{3}\right)^j\right] - \prod_{j=1}^{\infty}\left[1 - \left(\frac{1}{3}\right)^j\right] \propto$$

$$\propto 1 - \prod_{j=n+1}^{\infty}\left[1 - \left(\frac{1}{3}\right)^j\right] \propto 1 - \left(\frac{1}{3}\right)^n \quad \text{for} \quad n \to \infty \,. \tag{3.95}$$

Via $l_n \propto (1/2)^n$ for $n \to \infty$ then follows:

$$\mu(l) - \mu(0) \propto l^\beta \quad \text{with} \quad \beta = \log 3/\log 2 \,. \tag{3.96}$$

Let us now come back to the physical meaning of eq. (3.94). It answers both questions raised above. The measure $\mu(0)$ gives the fraction of chaotic parameter values in $r_\infty < r < 4$. The exponent β determines the probability p that a variation in r will change the qualitative behavior of the iterates. If one is sitting on a chaotic parameter value, p is proportional to the probability of finding a nonchaotic hole of size l, that is, $p \propto \mu(l) - \mu(0) \propto l^\beta$.

This situation means, for numerical computations of the logistic map (which are usually done with a precision $l \sim 10^{-14}$), that the odds of a mistake (i.e. that a trajectory believed to be chaotic is actually periodic) are, for $\beta \cong 0.45$, of the order 10^{-6}, which is acceptable.

According to Farmer (1985), one speaks only then of sensitive dependence on parameters, if $\beta < 1$ (i.e. if the odds of a mistake are larger than in the trivial case where one has $p \sim l^{-1}$).

It has been found (Farmer, 1986) that the set of parameter values where quasiperiodic behavior occurs in the subcritical circle map is also a fat fractal (see Chapter 6). This implies sensitive dependence on parameters which distinguish between quasi-periodic and mode-locked behavior (i.e. sensitive parameter dependence is not necessarily tied to chaos). Let us finally note that the fact that the exponent β is numerically the same for the logistic map $f(x)$ and the sine map $g(x)$ indicates a sort of *global universality* which is different from that originally found by Feigenbaum since it *applies to a set of positive measure* (volume) rather than just special points as period-doubling transitions.

Structural Universality

Structural universality in unimodal maps was discovered by Metropolis, Stein and Stein (1973). They considered the iterates of the logistic map $f(x)$ in the periodic windows. Starting from $x_0 = 1/2$, i.e. from the x value which corresponds to the maximum of $f(x)$, the sequence of iterates $f^n(x_0)$ on a periodic attractor can be characterized by a string $RL \ldots$ where R or L indicates whether $f^n(x_0)$ is to the right or left of x_0 (see e.g. Fig. 42).

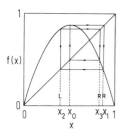

Fig. 42: The map $x_{n+1} = rx_n(1 - x_n)$ with $r = 3.49856$ displays a 4-cycle of the type *RLR*.

Table 3, which has been computed by Metropolis et al. (1973), shows that the sequence of strings is (up to cycles of length 7) the same for $f(x) = rx(1 - x)$ and $g(x) =$

Table 3: Universal sequences for two unimodal maps.

Period	U-sequence	Parameter value r in $x_{n+1} = rx_n(1 - x_n)$	Parameter value q in $x_{n+1} = q \sin(\pi x_n)$
2	R	3.2360680	.7777338
4	RLR	3.4985617	.8463822
6	RLR^3	3.6275575	.8811406
7	RLR^4	3.7017692	.9004906
5	RLR^2	3.7389149	.9109230
7	RLR^2LR	3.7742142	.9213346
3	RL	3.8318741	.9390431
6	RL^2RL	3.8445688	.9435875
7	RL^2RLR	3.8860459	.9568445
5	RL^2R	3.9057065	.9633656
7	RL^2R^3	3.9221934	.9687826
6	RL^2R^2	3.9375364	.9735656
7	RL^2R^2L	3.9510322	.9782512
4	RL^2	3.9602701	.9820353
7	RL^3RL	3.9689769	.9857811
6	RL^3R	3.9777664	.9892022
7	RL^3R^2	3.9847476	.9919145
5	RL^3	3.9902670	.9944717
7	RL^4R	3.9945378	.9966609
6	RL^4	3.9975831	.9982647
7	RL^5	3.9993971	.9994507

$q \sin(\pi x)$. This numerical result (which has actually been calculated for cycles up to length 11 and for other unimodal maps) suggests that the ordering of the sequence of $RL \ldots$ strings is universal for all maps on the [0, 1] interval which have a differentiable maximum and fall off monotonically on both sides. This so-called *structural universality* has been put on a rigorous footing by Guckenheimer (1980). It does not depend on the order of the maximum as the *metric universality* of Feigenbaum (1978) (a "metric" is needed there to measure the distances which scale). But it should be noted that structural universality seems to be restricted to one-dimensional maps because in higher dimensions, up to now, no ordering has been found, and one can have coexisting cycles of different length with different basins of attraction (see also sect. 5.7).

From the mathematical point of view, the sequence of cycles in an unimodal map $f(x)$ is completely described by Sarkovskii's theorem (1964). It states that if $f(x)$ has a point x which leads to a cycle of period p then it must also have a point x' which leads to a q-cycle for every $q \leftarrow p$ where q and p are elements in the following sequence:

$$
\begin{aligned}
&1 \leftarrow 2 \leftarrow 4 \leftarrow 8 \leftarrow 16 \ldots 2^m \ldots \leftarrow \\
&\ldots 2^m \cdot 9 \leftarrow 2^m \cdot 7 \leftarrow 2^m \cdot 5 \leftarrow 2^m \cdot 3 \ldots \leftarrow \\
&\ldots 2^2 \cdot 9 \leftarrow 2^2 \cdot 7 \leftarrow 2^2 \cdot 5 \leftarrow 2^2 \cdot 3 \ldots \leftarrow \\
&\ldots 2 \cdot 9 \leftarrow 2 \cdot 7 \leftarrow 2 \cdot 5 \leftarrow 2 \cdot 3 \ldots \leftarrow \\
&\ldots \quad 9 \leftarrow \quad 7 \leftarrow \quad 5 \leftarrow \quad 3
\end{aligned}
\tag{3.97}
$$

where the symbol $\leftarrow$ means "precede" (for a proof, see references of this chapter). It should be emphasized that Sarkovskii's theorem is only a statement concerning different x values at a fixed parameter value. It says nothing about the stability of the periods nor about the range of parameter values for which it could be observed. It follows, from the sequence in eq. (3.97), that if $f(x)$ has period three, then this implies that it must also have all periods n where n is an arbitrary integer. This is the famous theorem of Li and Yorke (1975) "Period three implies chaos". But it should be noted that "chaos" in this theorem means only aperiodic behavior and does not imply automatically a positive Liapunov exponent.

Chaotic Bands and Scaling

The logistic map at $r = 4$,

$$x_{n+1} = 4 x_n (1 - x_n) \equiv f_4(x_n),$$

(3.98)

can actually be solved by the simple change of variables:

$$x_n = \frac{1}{2} [1 - \cos(2\pi y_n)] \equiv h(y_n).$$

(3.99)

Then eq. (3.98) can be converted into

$$\frac{1}{2} [1 - \cos(2\pi y_{n+1})] = [1 - \cos(2\pi y_n)] [1 + \cos(2\pi y_n)] =$$

$$\frac{1}{2} [1 - \cos(4\pi y_n)]$$

(3.100)

which has one solution:

$$y_{n+1} = 2 y_n \bmod 1 \equiv g(y_n) \quad \text{or} \quad y_n = 2^n y_0 \bmod 1.$$

(3.101)

This implies the following solution to eq. (3.99):

$$x_n = \frac{1}{2} [1 - \cos(2\pi 2^n y_0)]$$

(3.102)

where $y_0 = \frac{1}{2\pi} \arccos(1 - 2x_0).$

(3.103)

Using eqns. (3.98–3.101), the invariant density $\rho_4(x)$ of $f_4(x)$ can be calculated from its definition:

$$\rho_4(x) = \lim_{N \to \infty} \frac{1}{N} \sum_{n-0}^{N-1} \delta(x - x_n) = \lim_{N \to \infty} \frac{1}{N} \sum_{n=0}^{N-1} \delta[x - h(y_n)] \; .$$

Using $\rho(y) = 1$ (which holds in analogy to the triangular map on page 30 also for the map in eq. (3.101)) eq. (3.102) becomes:

$$\rho_4(x) = \int_0^1 dy\, \rho(y)\, \delta[x - h(y)] = \frac{2}{|h'[y(x)]|} \tag{3.105}$$

i.e.

$$\rho_4(x = \frac{1}{\pi} \frac{1}{\sqrt{x(1-x)}} \tag{3.106}$$

as depicted in Fig. 43.

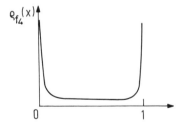

Fig. 43: The invariant density for $f_4 = 4x(1-x)$ (schematically)

These results show that the map $f_r(x)$ becomes ergodic for $r = 4$ and that the invariant density of a chaotic map need not always be a constant.
For the Liapunov exponent eq. (3.106) yields at $r = 4$:

$$\lambda = \int_0^1 dx\, \rho_4(x)\, |f'(x)| = \log 2 \tag{3.107}$$

i.e. the same value as for the map in eq. 3.101 which demonstrates that the Liapunov exponent is indeed invariant under a change of the coordinates.
Fig. 44 makes it plausible that the r-values for the inverse cascade (in which the chaotic regime at $r = 4$, which extends from $0 \le x \le 1$, is decomposed into finer and finer subintervals I_n that merge into the Feigenbaum attractor) are again determined by the law of functional composition.

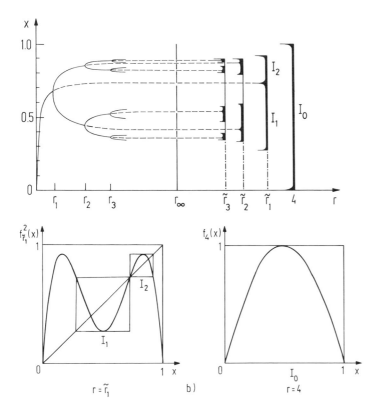

Fig. 44: a) Bifurcations for $r < r_\infty$ and the corresponding merging of chaotic regions for $r > r_\infty$; the dark areas indicate the corresponding invariant densities (see Fig. 43). Note the nonlinear scale on the abscissa. (After Grossmann and Thomae, 1977.) b) The lengths I_n of the chaotic intervals are again related to functional composition and the $\bar{r}_n$'s therefore scale like $\tilde{r}_n - r_\infty \sim \delta^{-n}$.

3.5 Parallels between Period Doubling and Phase Transitions

In the first part of this section we present a dictionary of the corresponding terms used in the bifurcation route to chaos and in the renormalization-group theory for second-order phase transitions. In the second part we summarize the measurable properties that characterize the Feigenbaum route and discuss some representative experiments.

We have already seen in Chapter 2 that the Liapunov exponent corresponds to the order parameter near a second-order phase transition. Table 3 shows that for the bifur-

---

cation route to chaos the analogy to a magnetic phase transition can be worked out in more detail. Both phenomena show a certain self-similarity (in the bifurcation pattern and in the pattern of spin-up/spin-down clusters near a critical point) which forms the basis for a renormalization-group treatment. Universality emerges then because there are only a few relevant eigenvalues (see also Appendices E and D).

We can also derive scaling laws for the Liapunov exponent λ and the correlation function $C(m)$ which are similar to those for the magnetization and the spin-spin correlation near a magnetic phase transition.

According to (2.9), the Liapunov exponent of a map f is (for $x_0 = 0$) is given by

$$\lambda(f) = \lim_{n \to \infty} \frac{1}{n} \sum_{i=0}^{n} \log |f'[f^i(0)]| \ . \tag{3.108}$$

Using

$$Tf \cdot Tf \cdots Tf = (Tf)^i = -\alpha f^{2^i}\left(-\frac{x}{\alpha}\right) \tag{3.109}$$

and

$$\frac{\mathrm{d}}{\mathrm{d}x} Tf = f'\left[f\left(-\frac{x}{\alpha}\right)\right] f'\left(\frac{x}{\alpha}\right) \tag{3.110}$$

we find

$$\lambda[Tf] = 2 \lim_{n \to \infty} \frac{1}{2n} \sum_{i=0}^{2n+1} \log |f'[f^i(0)]| = 2\lambda[f] \tag{3.111}$$

which can be iterated to

$$\lambda[f] = 2^{-n} \lambda[T^n f] \ . \tag{3.112]}$$

By choosing $f = f_R$, we can use

$$T^n f_R(x) = g(x) + (R - R_\infty) \cdot \delta^n \cdot a \cdot h(x) \tag{3.113}$$

in (3.112) which yields, by setting $(R - R_\infty)\delta^n = 1$, the scaling relation

$$\lambda_{f_R} = (R - R_\infty)^\beta \lambda[g(x) + a \cdot h(x)] \tag{3.114}$$

with $\beta = \log 2/\log \delta$ as a critical exponent.

This equation describes the approach of the Liapunov exponent to zero if a sequence of R's with the same μ (see Fig. 30) approaches R_∞; i.e., the power law $\lambda \propto (R - R_\infty)^\beta$ holds for the envelope of λ.

Table 4: Parallels between phase transitions and period doubling.

Phase transitions	Period doubling
Ginzburg-Landau Functional	One-dimensional map
$H = \int d^d x\,[c(\nabla\sigma)^2 + t\sigma^2 + u\sigma^4]$	$f_R(x)$
with parameter vector $\mu = (c,\,t,\,u)$	
Distance to the critical point	Distance from R_∞
$t = T - T_c$	$R - R_\infty$
Order parameter	Liapunov exponent
$\langle\sigma(x)\rangle$ (Magnetization)	λ_R (changes sign at R_∞)
Formation of block spins $\rightarrow$ renormalization-group transformation R with fixed point H^* ($\triangleq \mu^*$)	Functional composition $\rightarrow$ doubling operator T with fixed point g
$R[\mu^*] = \mu^*$	$T[g] = g$
Linearized renormalization-group transformation	Linearized doubling transformation
$R_{2^n}[\mu^*] = \mu^* + (T - T_c)\,2^{ny_1}\,\vec{e}_1$	$T^n f_R(x) = g(x) + (R - R_\infty)\cdot\delta^n\cdot a\cdot h(x)$
Parameter space:	Space of functions:

$\mu(t)$	critical surface	$f_R(x)$	stable manifold
		$h(x)$	
		$g(x)$	

| $\vec{e}_1$ = unstable direction | $a\cdot h(x)$ = one-dimensional unstable manifold |

In a similar way, for the correlation function (2.35)

$$C[m, f] = \lim_{n \to \infty} \frac{1}{n} \sum_{i=0}^{n} f^i(0) f^{i+m}(0) \qquad (3.115)$$

one finds the scaling relation

$$C[m, \mathrm{T}f] = \alpha^2 \lim_{n \to \infty} \frac{1}{n} \sum_{i=0}^{n} f^{2i}(0) f^{2i+2m}(0) =$$

$$= \alpha^2 \lim_{n \to \infty} \left\{ 2 \frac{1}{2n} \sum_{i=0}^{2n} f^i(0) f^{i+2m}(0) - \right. \qquad (3.116)$$

$$\left. - \frac{1}{n} \sum_{i=0}^{n-1} f^{2i}[f(0)] f^{2i+2m}[f(0)] \right\}$$

ie.

$$C[m, \mathrm{T}f] = \alpha^2 C[2m, f] \qquad (3.117)$$

and by using again (3.109):

$$C[m, f_R] = \alpha^{-2n} C[2^{-n} m, g(x) + (R - R_\infty) \cdot \delta^n \cdot a \cdot h(x)] . \qquad (3.118)$$

Eq. (3.118) leads to a variety of scaling laws, depending on which combination of variables we set equal to unity. We mention that at R_∞ the correlation function decays with a power law in m:

$$C[m, f_{R_\infty}] = \alpha^{-2n} C[2^{-n} m, g(x)] = m^{-\eta} C[1, g(x)] \qquad (3.119)$$

with $\eta = \log \alpha^2 / \log 2$.

These power laws have the following counterparts in magnetic phase transitions:

$$\lambda \propto |R - R_\infty|^\beta \qquad \triangleq M \propto |T - T_c|^\beta \qquad (3.120\,\mathrm{a})$$

$$C(m) \propto m^{-\eta} \quad \text{at} \quad R_\infty \triangleq C(|x|) \propto |x|^{-\eta} \quad \text{at} \quad T_c \qquad (3.120\,\mathrm{b})$$

where M is the magnetization and $C(|x|)$ is the spin-spin correlation function.

3.6 Experimental Support for the Bifurcation Route

After a preponderance of theory, let us now present some experimental support for the Feigenbaum route. First, we summarize its measurable fingerprints:

— There exists an infinite cascade of period doublings which leads to subharmonics in the power spectrum at frequencies $2^{-n} \cdot f_0$, where f_0 is the basic frequency.

— Each subharmonic lies below the preceeding level by a factor $\mu^{-1} = 0.1525$ $(10 \log_{10} \mu = 8.17$ dB$)$.

— The control parameter r scales for subsequent subharmonics n like $r_n - r_\infty \propto \delta^{-n}$.

— External noise destroys the fine structure of the power spectrum, and the noise level must decrease by a factor μ^{-1} to make one more subharmonic observable.

— The Poincaré map of the system is one-dimensional and shows a single quadratic maximum.

Following Feigenbaum's work, the bifurcation route to chaos has been found in many experimental systems, from the kicked pendulum and chemical reactions ... to optically bistable devices. Below we discuss three representative examples in more detail.

Figs. 45 and 46 show the power spectra for a Bénard experiment and for a nonlinear driven electrical RCL-oscillator.

The experimental set up for the Bénard experiment has already been described in Chapter 1. (We note that depending on the parameters of the liquid, the size of the cell, etc., the Bénard system exhibits different routes to chaos.) Libchaber and Maurer (1980) found the following properties of the Feigenbaum route in a Bénard experiment with liquid helium:

a) With increasing temperature difference (which is proportional to the control parameter r) there appear subharmonics of frequencies $f/2, f/4, f/8$ and $/16$ where f is the basic frequency.

b) Subsequent subharmonics differ by about 10 dB in qualitative agreement with theory ($\mu \triangleq 8.2$ dB).

Higher subharmonics are probably suppressed by external noise.

Although these results leave little doubt that the Feigenbaum route is involved, the explicit reduction of the hydrodynamic equations which describe the system to a one-dimensional Poincaré map with a single quadratic maximum has still not been demonstrated.

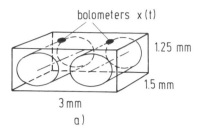

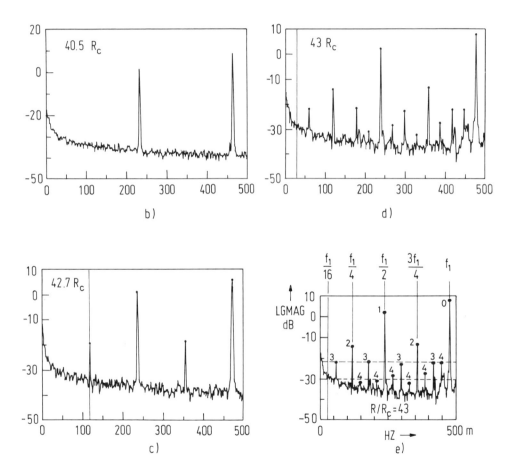

Fig. 45: a) Bénard cell with only two-roll convection pattern of liquid helium, basic frequency $f = 0.5 \text{ sec}^{-1}$. b)–d) Power spectrum of the temperature $x(t)$ with increasing Rayleigh number which is proportional to r. e) The heights of the nth subharmonics are compared with Feigenbaum's theory (horizontal lines). (After Libchaber and Maurer, 1980.)

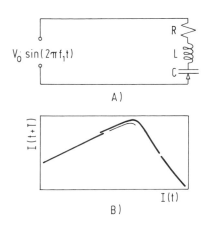

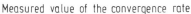

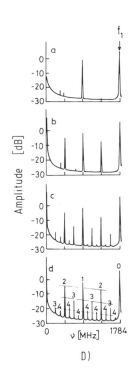

A)

B)

Measured value of the convergence rate

Subharmonic	$\Delta V_{threshold}$ (V)	δ_n
$f_1/2$		
	3.2 ± 0.02	
$f_1/4$		4.4 ± 0.1
	0.72 ± 0.02	
$f_1/8$		4.5 ± 0.6
	0.16 ± 0.02	
$f_1/16$		

C)

D)

Fig. 46: A) Circuit for the driven nonlinear RCL-oscillator. B) The observed current $I(t + T)$ vs. $T(t)$ yields a one-dimensional map with a single maximum. C) Determination of δ from the values of the control parameter V_0. D) a–c: Subharmonics in the power spectrum for increasing V_0; d: comparison with Feigenbaum's theory (horizontal lines). (After Linsay, 1981.)

The situation is somewhat better for the nonlinear *RCL*-oscillator shown in Fig. 46. The nonlinear element in this circuit is, according to Linsay (1981), the capacitor-diode, which leads to the following nonlinear relation between charge q and voltage V:

$$V(g) = \left[1 + \frac{V(q)}{0.6} \right]^{0.43} \frac{q}{C_0} . \qquad (3.121)$$

The differential equation for the time dependence of q is

$$L\ddot{q} + R\dot{q} + V(q) = V_0 \sin(2\pi f_1 t) \qquad (3.122)$$

and the circuit acts like an analog computer for a driven nonlinear oscillator. Fig. 46 shows that for special values of V_0 (which is proportional to the control parameter r) the sequence of current signals $I_n = I(t_0 + nT)$, where the time $T = 1/f_1$, can indeed be generated from a one-dimensional map with a quadratic maximum. (The current is related to the charge via $I = \dot{q}$, and I_n corresponds to x_n). The corresponding power

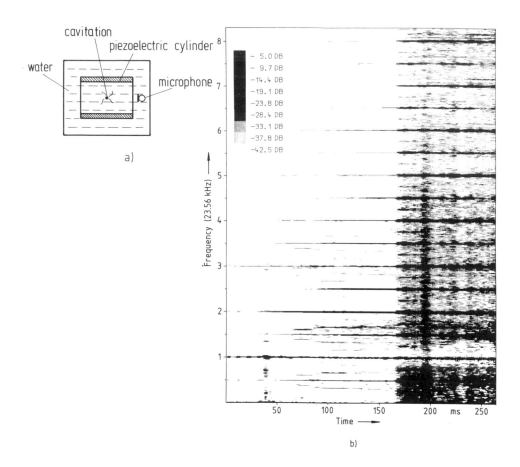

a)

b)

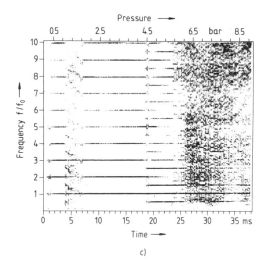

c)

Fig. 47: Experimental set up for generation and detection of cavitation noise (a). Sequence of observed (b) and calculated (c) power spectra for different input pressures. The noise amplitude is encoded as grey scale and the input pressure (which is measured experimentally by the voltage at the driving piezoelectric cyclinder) is increased linearly in time. See also the colored version of c) in Plate V at the beginning of the book. (After Lauterborn and Cramer, 1981.)

spectrum exhibits — as expected — all the features of the bifurcation route and yields an estimate for δ which deviates by 10% from Feigenbaum's asymptotic value. See also the phase portraits ($I(t)$ versus $V(t)$) for the nonlinear RCL-oscillator of Lauterborn et al. (1984) in Plate I at the beginning of the book. We note also that there exists theoretical (Rollins and Hunt, 1982) and experimental (S. Martin, priv. comm.) evidence that the chaotic behavior of RCL-oscillators with Varactor diodes (which were used in the experiments above) is not caused by the nonlinearity of the diode but by its large recovery time. But, this situation can again be described by a one-dimensional noninvertible map.

To demonstrate that the Feigenbaum route indeed occurs in quite different systems, we finally describe an experiment by Lauterborn and Cramer (1981) in which this route has been observed in acoustics (Fig. 47a). They irradiated water with sound of high intensity and measured the sound output of the liquid. The nonlinear elements in this system are cavitations, i.e. bubbles filled with water vapor which are created by the pressure gradients of the initial sound wave and whose wall oscillations are highly nonlinear.

Fig. 47 shows a sequence of power spectra that is obtained experimentally (b) and from a numerical calculation (c) (in which only a single spherical bubble was considered). With increasing input pressure (which is the external control parameter), one observes a subharmonic route to chaos that, besides the sequence $f_0 \rightarrow f_0/2 \rightarrow f_0/4 \ldots$ $\rightarrow$ chaos, also contains $f_0/3$. Moreover, the system shows signs of reverse bifurcations where it returns from chaotic behavior to a line spectrum.

4 The Intermittency Route to Chaos

By intermittency we mean the occurrence of a signal that alternates randomly between long regular (laminar) phases (so called intermissions) and relatively short irregular bursts. Such signals have been detected in a large number of experiments. It has also been observed that the number of chaotic bursts increases with an external parameter, which means that intermittency offers a continuous route from regular to chaotic motion.

In the first section of this chapter, we present mechanisms for this phenomenon proposed by Pomeau and Manneville (1979) and discuss type-I intermittency which is generated by an inverse tangent bifurcation. It is shown in the second section that the transition to chaos via intermittency has in fact universal properties and represents one of the rare examples where the (linearized) renormalization-group equations can be solved exactly. These results will be used in Section 3 to demonstrate that intermittency provides a universal mechanism for $1/f$-noise in nonlinear systems. In the final section, we summarize typical properties of the intermittency route and discuss some experiments.

4.1 Mechanisms for Intermittency

The intermittency route to chaos has been investigated in a pioneering study by Pomeau and Manneville (1979). They solved numerically the differential equations of the Lorenz model,

$$\dot{X} = \sigma(X - Z) \tag{4.1a}$$

$$\dot{Y} = -XZ + rX - Y \tag{4.1b}$$

$$\dot{Z} = XY - bZ \tag{4.1c}$$

and for the Y-component they found the behavior shown in Fig. 48.

For $r < r_c$, $Y(t)$ executes a stable periodic motion. Above the threshold r_c, the oscillations are interrupted by chaotic bursts, which become more frequent as r is increased until the motion becomes truly chaotic.

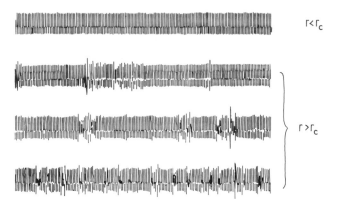

$r < r_c$

$r > r_c$

Fig. 48: Time plot of one coordinate in the Lorenz model (after Pomeau and Manneville 1980).

Pomeau and Manneville gave the following interpretation for this behavior:

The stable oscillations for $r < r_c$ correspond to a stable fixed point in the Pioncaré map (see also Fig. 6). Above r_c this fixed point becomes unstable. Because there are essentially three ways in which a fixed point can loose its stability (in all of them the modulus of the eigenvalues of the linearized Poincaré map becomes larger than unity), Pomeau and Manneville distinguished the three types of intermittency shown in Table 5. (See also Table 7, p. 98, for the form of the signal.)

Type-I Intermittency

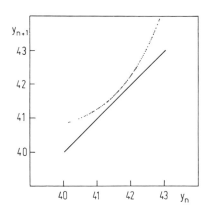

Fig. 49: Poincaré map of the Lorenz model for r slightly above $r_c = 166$ (after Pomeau and Manneville, 1980).

Fig. 49 shows a Poincaré map for the Lorenz model, after Pomeau and Manneville who plotted the values y_n where $y(t)$ crossed the plane $x = 0$. If this figure is compared with Table 5, it is seen that the Lorenz model displays intermittency of type I.

This transition to chaos is charaterized by an *inverse tangent bifurcation* in which two fixed points (a stable and an unstable one) merge as depicted in Fig. 50.

4   The Intermittency Route to Chaos

Table 5: Three types of intermittency.

Type	Charateristic behavior and maps	Typical map ($\varepsilon < 0 \rightarrow \varepsilon > 0$)	Eigenvalues
I	A real eigenvalue crosses the unit circle at $+1$ $$x_{n+1} = \varepsilon + x_n + u x_n^2$$		
II	Two conjugate complex eigenvalues cross the unit circle simultaneously. $$r_{n+1} = (1 + \varepsilon) r_n + u r_n^3$$ $$\theta_{n+1} = \theta_n + \Omega$$		
III	A real eigenvalue crosses the unit circle at -1 $$x_{n+1} = -(1 + \varepsilon) x_n - u x_n^3$$		

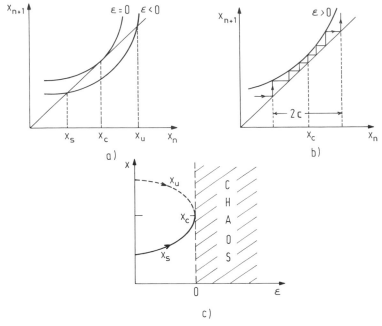

Fig. 50: Mechanism for type-I intermittency. a) Poincaré map for $\varepsilon = r - r_c \leq 0$, b) Poincaré map for $\varepsilon > 0$ and motion of the trajectory, (note that the "ghost of the fixed point" x_c attracts trajectories on the left hand side and repells them on the right hand side), c) inverse tangent bifurcation.

For $r > r_c$, the map has no stable fixed points. However, a sort of "memory" of a fixed point is displayed since the motion of the trajectory slows down in the vicinity of x_c, and numerous iterations are required to move through the narrow channel between the map and, the bisector. This leads to the long laminar regions for values of r just above r_c in Fig. 48. After the trajectory has left the channel, the motion becomes chaotic until reinjection into the vicinity of x_c starts a new regular phase. The theory of Pomeau and Manneville explains only the laminar motion but gives no information about the mechanism which generates chaos.

Another example for type-I intermittency appears in the logistic map

$$x_{n+1} = f_r(x_n = rx_n(1 - x_n) . \tag{4.2}$$

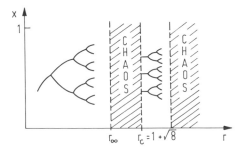

Fig. 51: Window with period three in the chaotic regime.

Numerically, it is found that for $r_c = 1 + \sqrt{8}$ this map exhibits a cycle of period three with subsequent bifurcations, i.e., there is a window in the chaotic regime as shown schematically in Fig. 51. The iterates for r-values larger and smaller than r_c are shown in Fig. 52. There is a regular cycle of period three slightly above r_c; but below r_c, laminar regions occur interrupted by chaos.

An explanation of this peculiar behavior follows from Fig. 53, which shows the third iterate of $f_r(x)$ at $r = r_c$. There are three fixed points that become unstable for $r < r_c$ and lead to intermittency of type I. It should be noted that inverse tangent bifurcations provide (in contrast to pitchfork bifurcations in which the number of fixed points is doubled) the only mechanism by which an uneven number of fixed points can be generated in the logistic map.

Length of the Laminar Region

As the next step, we calculate the average length $\langle l \rangle$ of a laminar region (as a function of the distance $\varepsilon = r - r_c$ from the critical point) for the logistic map. It will become clear from our derivation that the result for $\langle l \rangle (\varepsilon)$ is not confined to this special map but holds for any Poincaré map that leads to type-I intermittency.

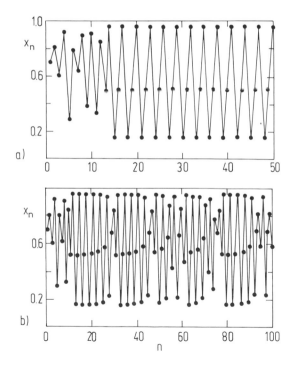

Fig. 52: Iterates of the logistic map starting from $x = 0.7$; a) in the stable three-cycle region $r_c - r = -0.02$; b) in the intermittent region $r_c - r = 0.002$. (After Hirsch et al. 1981.)

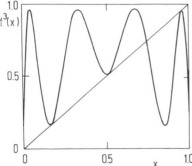

Fig. 53: The threefold iterated map $f_r^3(x)$ for $r = r_c$.

Expanding $f_r^3(x)$ around the points x_c and r_c that are determined by

$$\frac{d}{dx} f_{r_c}^3(x_c) = 1, \quad f_{r_c}^3(x_c) = x_c \tag{4.3}$$

we obtain

$$f_r^3(x) = f_r^3[x_c + (x - x_c)] = x_c + (x - x_c) + a_c (x - x_c)^2 + b_c (r - r_c) \tag{4.4}$$

where

$$2a_c \equiv \left. \frac{d^2 f_r^3}{dx^2} \right|_{x_c, \, r_c} \qquad\qquad b_c \equiv \left. \frac{d f_r^3}{dr} \right|_{x_c, \, r_c} \tag{4.5}$$

(A similar equation holds for all three fixed points of $f_{r_c}^3(x)$, but we choose the middle point for convenience the result is independent of the constants).

With $y \equiv (x - x_c)/b_c$ and $a \equiv a_c \cdot b_c > 0$ the recursion for the map

$$x_{n+1} = f_r^3(x_n) \tag{4.6}$$

transform via (4.4) in the vicinity of x_c into

$$y_{n+1} = y_n + a y_n^2 + \varepsilon; \quad \varepsilon = r - r_c . \tag{4.7}$$

(A similar map is obtained if the Poincaré map in Fig. 49 is expanded around the point of tangency). The laminar regions are now defined by the requirement that subsequent iterates change only very little; i.e., their distance to x_c should be smaller than a threshold value c:

$$|y_n| < c \ll 1 . \tag{4.8}$$

---

4   The Intermittency Route to Chaos

In this region we can therefore safely replace the difference equation (4.7) by the differential equation,

$$\frac{dy}{dl} = ay^2 + \varepsilon \tag{4.9}$$

(l counts the iterations in the laminar region) which after integration yields

$$l(y_{out}, y_{in}) = \frac{1}{\sqrt{a\varepsilon}} \left[\arctan\left[\frac{y_{out}}{\sqrt{\varepsilon/a}} \right] - \arctan\left[\frac{y_{in}}{\sqrt{\varepsilon/a}} \right] \right] \tag{4.10}$$

To find the average length $\langle l \rangle$ of a laminar region, we assume that after having left the laminar region at $y_{out} = c$, the point becomes, after some irregular bursts, reinjected to $|y| < c$ at y_{in} with a probability function $P(y_{in})$, which is symmetric about x_c, i.e. $P(y_{in}) = P(-y_{in})$.
This yields

$$\langle l \rangle = \int_{-c}^{c} dy_{in} P(y_{in}) l(c, y_{in}) = \frac{1}{\sqrt{a\varepsilon}} \arctan\left[\frac{c}{\sqrt{\varepsilon/a}} \right]. \tag{4.11}$$

For $c/\sqrt{\varepsilon/a} \gg 1$, the average length $\langle l \rangle$ varies as

$$\langle l \rangle \propto \varepsilon^{-1/2} . \tag{4.12}$$

This characteristic variation was first derived by Pomeau and Manneville (1980) and is valid numerically for the logistic map as shown in Fig. 54.

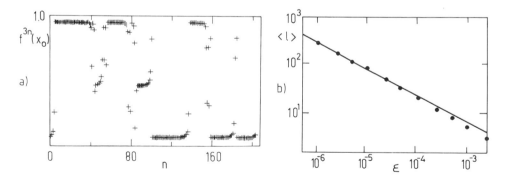

Fig. 54: a) Sequence of third iterates $f_r^{3n}(x_0$ for $r_c - r = 0.0001$ showing regions of laminar behavior interrupted by intermittent irregularities; b) the points show $\langle l \rangle$ versus ε for $c = 10^{-2}$, and the line is the asymptotic limit $\langle l \rangle = (\pi/2)(\varepsilon a)^{-1/2}$. (After Hirsch et al., 1981.)

4.2 Renormalization-Group Treatment of Intermittency

The intermittency phenomenon has also been investigated by the renormalization-group method using the doubling operator which we encountered previously for the Feigenbaum route. The idea is as follows: One considers a generalization $f(x)$ of the map (4.7) for $\varepsilon = 0$ to arbitrary exponents $z > 1$ which for $x \to 0$ has the form

$$f(x \to 0) = x + u|x|^z . \tag{4.13}$$

Its second iterate $f^2(x)$ shows (because of the linear term in x), after proper rescaling, the same asymptotic behavior (see Fig. 55). This is reminiscent of Fig. 29 for the logistic map. It could, therefore, be asked whether repeated application of the doubling operator T to a function of type (4.13) could also lead to a fixed point $f^*(x)$ of T :

$$Tf^*(x) = \alpha f^* \left[f^* \left(\frac{x}{\alpha} \right) \right] = f^*(x) \tag{4.14}$$

but with the boundary conditions (4.13), i.e. $f^*(0) = 0$ and $f^{*\prime}(0) = 1$ instead of $f^*(0) = 1$ and $f^{*\prime}(0) = 0$ for the Feigenbaum bifurcations.

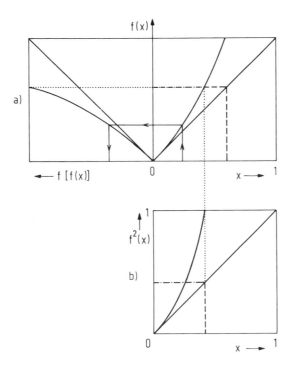

Fig. 55: $f(x)$ and $f^2(x)$ vs. x for $x \geq 0$. a) The second iterate is constructed by plotting, in addition to the square which contains $f(x)$, the same square rotated through 90°: The arrows indicate how $f[f(x)]$ is obtained. b) The second iterate $f^2(x)$ looks similar to the original $f(x)$ in the dotted square.

It has been shown by Hu and Rudnick (1982) that together with the new boundary condition (4.13) the fixed-point equation (4.14), which is characteristic for intermittency, can be solved exactly.

Here, the trick is to write the recursion relation

$$x' = f(x) \tag{4.15}$$

in implicit form

$$G(x') = G(x) - a \tag{4.16}$$

i.e.

$$x'(x) = G^{-1}[G(x) - a] = f(x) \tag{4.17}$$

where a is a free parameter.

The fixed-point equation

$$\alpha f^*[f^*(x)] = f^*(\alpha x) \tag{4.18}$$

then becomes

$$\alpha x''(x) = x'(\alpha x) \tag{4.19}$$

or by operating on this with G:

$$G(\alpha x'') = G[x'(\alpha x)] = G(\alpha x) - a . \tag{4.20}$$

Next, eq. (4.16) is used to obtain

$$G(x'') = G(x') - a = G(x) - 2a \tag{4.21}$$

i.e.

$$\frac{1}{2} G(x'') = \frac{1}{2} G(x) - a . \tag{4.22}$$

Comparison of (4.20) and (4.22) indicates that to solve the fixed-point equation, G must have the property

$$\frac{1}{2} G^*(x) = G^*(\alpha x) . \tag{4.23}$$

The simple choice $G^*(x) = |x|^{-(z-1)}$ with $\alpha = 2^{\frac{1}{(z-1)}}$ yields the desired result. The fixed-point function therefore becomes

$$f^*(x) = G^{*-1}[G^*(x) - a] = [|x|^{-(z-1)} - a]^{-1/(z-1)} \tag{4.24}$$

which for $a = (z-1)u$ fulfills boundary condition (4.13). This derivation shows that the fixed-point map for intermittency is mathematically related to a translation $G(x') = G(x) - a$; however, a simple physical explanation for this connection is not clear.

It is of course enough to find the fixed-point function $f^*(x)$ but one wants to classify the perturbations to f^* according to their relevance (see, e.g., Table 4). We investigate, therefore, how the doubling transformation T acts (to linear order in ε) on a function

$$f_\varepsilon(x) = f^*(x) + \varepsilon h_\lambda(x) \quad \text{for} \quad \varepsilon \ll 1 . \tag{4.25}$$

Using the definition (4.14) for T we find:

$$T f_\varepsilon = \alpha f_\varepsilon \left[f_\varepsilon \left(\frac{x}{\alpha} \right) \right] \tag{4.26}$$

$$= \alpha f^* \left[f^* \left(\frac{x}{\alpha} \right) + \varepsilon h_\lambda \left(\frac{x}{\alpha} \right) \right] + \varepsilon \alpha h_\lambda \left[f^* \left(\frac{x}{\alpha} \right) + \varepsilon h_\lambda \left(\frac{x}{\alpha} \right) \right]$$

$$= \alpha f^* \left[f^* \left(\frac{x}{\alpha} \right) \right]$$

$$+ \varepsilon \alpha \left\{ f^{*'} \left[f^* \left(\frac{x}{\alpha} \right) \right] h_\lambda \left(\frac{x}{\alpha} \right) + h_\lambda \left[f^* \left(\frac{x}{\alpha} \right) \right] \right\} + O(\varepsilon^2)$$

$$= f^*(x) + \lambda \varepsilon h_\lambda(x) + O(\varepsilon^2) .$$

The last equation holds only if $h_\lambda(x)$ is an eigenfunction, with the eigenvalue λ, of the linearized doubling operator L_{f^*}:

$$L_{f^*}[h_\lambda(x)] \equiv \alpha \{ f^{*'}[f^*(x)] h_\lambda(x) + h_\lambda[f^*(x)] \} = \lambda h_\lambda(\alpha x) \tag{4.27}$$

by analog to eq. (3.50) for the Feigenbaum route.

We now show that the method used above (to find the fixed-point function f^*) allows us also to find the spectrum of eigenvalues λ and the corresponding eigenfunctions h_λ.

First we write $f_\varepsilon(x)$ in implicit form using eq. (4.17):

$$f_\varepsilon(x) = f^*(x) + \varepsilon h_\lambda(x) = x' = G_\varepsilon^{-1}[G_\varepsilon(x) - a] . \tag{4.28}$$

If we expand

$$G_\varepsilon(x) = G^*(x) + \varepsilon H_\lambda(x) \tag{4.29}$$

then $h_\lambda(x)$ can be expressed in terms of $H_\lambda(x)$ (and vice versa) by comparing the factors linear in ε on both sides of (4.28).

Next we consider the second iterate,

$$x''(x) = f_\varepsilon[f_\varepsilon(x)] \tag{4.30}$$

and apply G_ε to this. This yields

$$G_\varepsilon(x'') = G_\varepsilon(x') - a = G_\varepsilon(x) - 2a \tag{4.31}$$

or more explicitly:

$$G^*(x'') + \varepsilon H_\lambda(x') = G^*(x) + \varepsilon H_\lambda(x) - 2a . \tag{4.32}$$

Because $G^*(x)$ has the form of a simple power of x we try a similar ansatz for $H_\lambda(x)$:

$$H_\lambda(x) = |x|^{-p} . \tag{4.33}$$

Using the property (4.23) of $G^*(x)$, (4.32) then becomes

$$G^*(\alpha x'') + \lambda \varepsilon H_\lambda(\alpha x'') = G^*(\alpha x) + \lambda \varepsilon H_\lambda(\alpha x) - a \tag{4.34}$$

or

$$G_{\lambda\varepsilon}(\alpha x'') = G_{\lambda\varepsilon}(\alpha x) - a \tag{4.35}$$

$$\rightarrow \alpha x'' = G_{\lambda\varepsilon}^{-1}[G_{\lambda\varepsilon}(\alpha x) - a] \tag{4.36}$$

where

$$\lambda = 2^{\frac{p+1-z}{z-1}} . \tag{4.37}$$

With (4.28) this translates into

$$\alpha f_\varepsilon[f_\varepsilon(x)] = f_{\lambda\varepsilon}(\alpha x) = f^*(\alpha x) + \lambda \varepsilon h_\lambda(\alpha x) . \tag{4.38}$$

By comparing this result with eq. (4.26) we see that λ is indeed the eigenvalue of h_λ, which is determined by

$$f^*(\alpha x) + \lambda \varepsilon h_\lambda(\alpha x) = G_{\lambda\varepsilon}^{-1}[G_{\lambda\varepsilon}(\alpha x) - a] . \tag{4.39}$$

Solving (4.39) to order ε one obtains

$$h_\lambda(x) = \frac{1}{up} \left[|x|^{-(z-1)} - u(z-1) \right]^{-\frac{z}{z-1}}$$
$$\cdot \left\{ |x|^{-p} - \left[|x|^{-(z-1)} - u(z-1) \right]^{-\frac{p}{z-1}} \right\} . \tag{4.40}$$

Eqns. (4.37) and (4.40) represent the main results of this section. They provide the information as to how T acts (to linear order in the deviation $f - f^*$) on a function f that obeys the boundary condition (4.13), because we obtain by expanding $f(x) - f^*(x)$ into $h_\lambda(x)$:

$$T^n f(x) = T^n(f^*(x) + f(x) - f^*(x)) \tag{4.41}$$
$$= T^n(f^*(x) + \sum_\lambda c_\lambda h_\lambda(x)) = f^*(x) + \sum_\lambda \lambda^n c_\lambda h_\lambda(x)$$

The intermittency route therefore represents — in contrast to the Feigenbaum route — one of the rare examples where the linearized renormalization-group equations can be solved exactly.

As an application, we now calculate the dependence of the duration $\langle l \rangle$ of a laminar region on the shift ε of the map from tangency shown in Fig. 56.

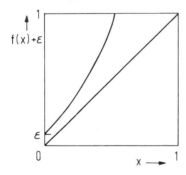

Fig. 56: Shift ε from tangency.

The eigenfunction h_λ in Eq. (4.40) has been normalized, so that its lowest order term in x is $|x|^{2z-1-p}$. We, therefore, see that a constant shift ε from tangency corresponds to a relevant perturbation with $p = 2z - 1$. The eigenvalue λ_ε that corresponds to this is

$$\lambda_\varepsilon = 2^{\frac{z}{z-1}} \tag{4.42}$$

With this information we can determine $\langle l \rangle(\varepsilon)$ by a simple scaling procedure.

Because $\langle l \rangle$ is related to the number of iterates of x_0, and $f^2(x) = f[f(x)]$ only requires half as many steps as $f(x)$, we arrive at the scaling relation

$$\langle l \rangle [Tf(x_0)] = \frac{1}{2} \langle l \rangle [f(x_0)] . \tag{4.43}$$

Using (4.26) this becomes after many iterations

$$\langle l \rangle \, [f(x_0)] \; = \; 2^n \, \langle l \rangle \, [T^n f(x_0) \; = \; f^*(x_0) \; + \; \varepsilon \lambda_\varepsilon^n h_\lambda (x_0)]$$

from which for $\varepsilon \lambda_\varepsilon^n \; = \; 1$ we obtain with (4.42):

$$\langle l \rangle \propto \varepsilon^{-\nu} \quad \text{with} \quad \nu = \frac{z-1}{z} \, . \tag{4.44}$$

For $z = 2$ this agrees with our previous result (4.12). One can show with the same method that a perturbation which is linear in x, i.e.

$$f(x) \; = \; f^*(x) \; + \; \varepsilon x \tag{4.45}$$

leads to

$$\langle l \rangle \propto \varepsilon^{-1} \tag{4.46}$$

and perturbations εx^m with $m > z$ are irrelevant.

Finally, we mention that the effect of external noise with amplitude σ on intermittency has been treated by Hirsch, Nauenberg, and Scalapino (1982) with the net result that $\langle l \rangle$ scales like

$$\langle l \rangle \; = \; \varepsilon^{-\nu} g \, (\sigma^\mu \varepsilon) \quad \text{with} \quad \mu = \frac{z-1}{z+1} \tag{4.47}$$

where g is a universal function.

4.3 Intermittency and $1/f$-Noise

It has been observed experimentally that the power spectra S_f of a large variety of physical systems (see Table 6) diverge at low frequencies with a power law $1/f^\delta$ ($0.8 < \delta < 1.4$). This phenomenon is called $1/f$-noise. Despite considerable theoretical efforts, a general theory encompassing $1/f^\delta$-divergencies in several experiments is still lacking.

In the following, we show that a class of maps which generates intermittent signals also displays $1/f^\delta$-noise, and we link the exponent δ to the universal properties of the map using the renormalization-group approach. Although the intermittency mechanism for $1/f$-noise is − as we shall demonstrate below − well verified numerically for maps, is still remains unresolved, whether it also provides an explanation for the experiments shown in Table 6. (We do not think that the intermittency mechanism which is very sensitive to external pertubations could explain the robust $1/f$-noise found in resistors. But there is a good chance to find this mechanism in chemical reactions and in the Bénard convection; see Manneville (1980), and Dubois et al. (1983).)

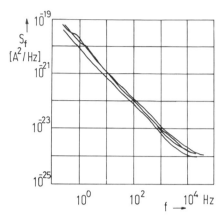

Fig. 57: Example for $1/f$-noise in the current of a bipolar transistor (after D. Wolf, 1978).

Table 6: Systems showing $1/f$-noise.

System	Signal
Carbon film	Current
Metal film	Current
Semiconductor	Current
Metal contact	Current
Semiconductor contact	Current
Ionic solution contact	Current
Superconductor	Flux flow
Vaccum tube	Current
Junction diode	Current
Schottky diode	Current
Zener diode	Current
Bipolar transistor	Current
Field effect transistor	Current
Thermocell	Thermovoltage
Electrolytic concentration cell	Voltage
Quartz oscillator	Frequency
Earth (5 days mean of rotation)	Frequency
Sound and speech sources	Loudness
Nerve membrane	Potential
Highway traffic	Current

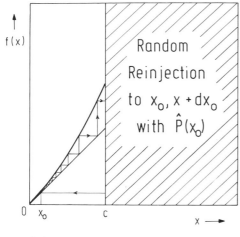

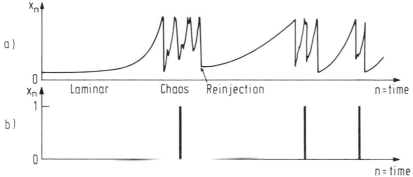

Fig. 58: The map $f(x)$ has the limiting behavior $f(x \to 0) = x + ux^z$ and is arbitrary beyond $x = c$ with the only requirement that this part of the map produces random reinjection into the region $0 \le x_0 \le c$ with a probability $\hat{P}(x_0)$.

Fig. 59: a) The iterates $x_n = f^n(x_0)$ as a function of time, showing laminar and chaotic behavior according to whether the trajectory is in $[0, c]$ or in the chaotic region; b) the idealized signal.

We want to calculate the power spectrum S_f for the map

$$x_{n+1} = f(x_n) \tag{4.48}$$

in Fig. 58 where $x_n \ge 0$. In other words, we only use that part of the map where the "ghost of the fixed point" is repulsive (compare Figs. 50 and 65). Therefore our mechanism for $1/f$-noise only works for type-III (and type-II) intermittency (Ben-Mizrachi et al., 1985). It is useful to express S_f via the correlation function $C(m)$:

$$S_f \propto \lim_{N \to \infty} \frac{1}{N} \sum_{m=0}^{N} \cos(2\pi mf) C(m) \tag{4.49}$$

where

$$C(m) = \lim_{N \to \infty} \frac{1}{N} \sum_{n=0}^{N} x_{n+m} x_n . \tag{4.50}$$

(This result follows by Fourier transformation from the definitions in eqns. (4.49) and (4.50).) To evaluate $C(m)$, we idealize the signal as shown in Fig. 59b; i.e., we assume that x_n is practically zero in the laminar regions and replace the short burst regions by lines of height one. $C(m)$ then becomes proportional to the conditional probability of finding a signal at time m, given that there occured a signal at time zero.

Next, we express $C(m)$ in terms of the probability $P(l)$ of finding an intermission of length l, which we shall calculate below in a universal way.

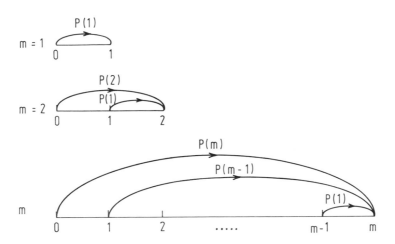

Fig. 60: The probability of finding a signal at m, assuming there was a signal at zero, can be expressed by $P(l)$.

Fig. 60 shows that

$$C(1) = P(1)$$
$$C(2) = P(2) + P(1)^2 = P(2) + C(1)P(1)$$
$$C(m) = P(m) + C(1)P(m-1) + \ldots + C(m-1)P(1) \qquad (4.51)$$

which can be written as

$$C(m) = \sum_{k=0}^{m} C(m-k)P(k) + \delta_{m,0} \qquad (4.52)$$

if we define $P(0) = 0$, $C(0) = 1$.

We now use eq. (4.24) to calculate the probability $P(l)$ of finding a laminar region of length l for (4.48).

$P(l)$ is related to the probability $\hat{P}(x_0)$ via

$$\hat{P}(x_0)\, dx_0 = \hat{P}[x_0(l)] \left| \frac{d x_0}{d l} \right| dl \equiv P(l)\, dl \qquad (4.53)$$

$$\rightarrow P\,(l) \;=\; \hat{P}\,[x_0\,(l)] \left| \frac{dx_0}{dl} \right| \tag{4.54}$$

since it follows from Fig. 58 that

$$f^l(x_0) \;=\; c \;\rightarrow\; x_0 \;=\; x_0\,(l) \tag{4.55}$$

$x_0\,(l)$ can be calculated by using the doubling operator. In the absence of relevant perturbations (which will be discussed later), we have

$$T^n f(x_0) \;=\; \alpha^n f^{2n}\,(x_0/\alpha^n) \;\approx\; f^*\,(x_0) \,, \quad \text{for} \quad n \gg 1 \tag{4.56}$$

i.e. the function is driven to the fixed point. This yields

$$f^{2n}\,(x_0) \;=\; \alpha^{-n} f^*\,(\alpha^n x_0) \,. \tag{4.57}$$

Here both $\alpha = 2^{\frac{1}{z-1}}$ and $f^*\,(x) = |x|\,[1 - (z-1)\,u\,|x|^{z-1}]^{-\frac{1}{z-1}}$ depend only on z which determines the universality class. If we use (4.57) in (4.55), we obtain for $l = 2^n$):

$$x_0\,(l) \;\propto\; l^{\frac{1}{z-1}} \tag{4.58}$$

and with (4.54) this yields the desired universal result for $P\,(l)$,

$$P\,(l) \;\propto\; \hat{P}\,(0)\,l^{-\frac{z}{z-1}} \,. \tag{4.59}$$

Here we assumed that $\hat{P}\,(x_0)$ varies only slowly with x_0, i.e. $\hat{P}\,(x_0 \propto l^{-\frac{1}{z-1}} \rightarrow 0)$ $\approx \hat{P}\,(0)$, for $l \gg 1$.

We pass on to continuous time variables since we are only interested in the long-time limit, and solve (4.52) by Laplace transformation using the convolution theorem. This yields

$$C_s \;=\; \frac{1}{1 - P_s} \quad \text{with} \quad g_s \equiv \int\limits_0^\infty dt\, e^{-st} g\,(t) \tag{4.60}$$

from which we obtain S_f as

$$S_f \;=\; \int\limits_0^\infty dt\, \cos\,(2\pi f t)\, C\,(t) \;=\; \frac{1}{2}\,[C_{s\rightarrow 2\pi if} + C_{s\rightarrow -2\pi if}] \,. \tag{4.61}$$

Substitution of $P\,(l)$ from (4.59) into (4.60) and (4.61) yields

$$\lim_{f \to 0} S_f \propto \begin{cases} f^{-\frac{2z-5}{z-1}} & z > 3 \\[2mm] \dfrac{|\log f|^2}{f^{1/2}} & z = 3 \\[2mm] f^{-\frac{1}{z-1}} & 2 < z < 3 \\[2mm] \dfrac{1}{f|\log f|^2} & z = 2 \\[2mm] f^{-\frac{2z-3}{z-1}} & \dfrac{3}{2} < z < 2 \\[2mm] |\log f| & z = \dfrac{3}{2} \\[2mm] \text{const.} & z < \dfrac{3}{2} \end{cases} \qquad (4.62)$$

(The results for $z \geq 3$ are from Ben-Mizrachi et al., 1985.)

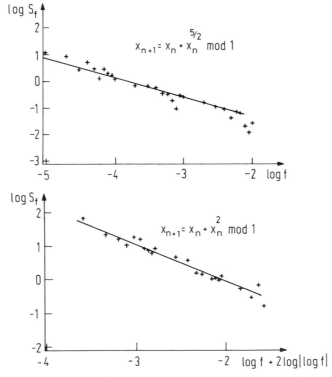

Fig. 61: Numerically determined power spectra for $z = 5/2$ and $z = 2$ compared to eq. (4.62) (after Procaccia and Schuster, 1983).

Fig. 61 shows that this result agrees reasonably well with the numerically determined power spectra of the map

$$x_{n+1} = x_n + x_n^z \bmod 1$$

for

$$z = \frac{5}{2} \quad \text{and} \quad z = 2 . \tag{4.63}$$

Let us now briefly discuss the effect of perturbations. The low frequency divergence of the power spectrum arises because arbitrarily long laminar regions ($P(l) \propto l^{-z/(z-1)}$) occur with finite probability in the (unperturbed) map in Fig. 58. But we also showed in Section 2 that in the presence of relevant perturbations (as e.g. a shift ε from tangency) the average duration of an intermission becomes finite:

$$\langle l \rangle \sim \varepsilon^{-\nu} . \tag{4.64}$$

This yields a cutoff

$$f_c \sim \langle l \rangle^{-1} \sim \varepsilon^\nu \tag{4.65}$$

in the $1/f^\delta$-behavior of S_f as shown in Fig. 62.

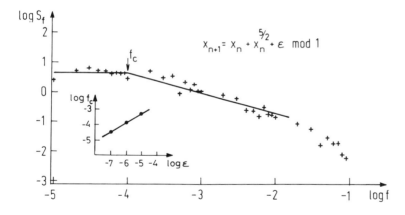

Fig. 62: Power spectrum of the map $x_{n+1} = \varepsilon + x_n + x_n^{5/2} \bmod 1$. The inset shows the scaling of f_c with ε as predicted in eq. (4.62) (after Procaccia and Schuster, 1983).

4.4 Experimental Observation of the Intermittency Route

Table 7 summarizes some measurable characteristic properties of the intermittency route to chaos. The different types of intermittency can be distinguished by the form of the signal and by the distribution $P(l)$ of the laminar lengths.

Table 7: Characteristic properties of different types of intermittency

Type	Poincaré map	Laminar Signal	Distribution $P(l)$
I	 $$x_{n+1} = x_n + x_n^2 + \varepsilon$$	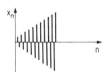 increases monotonously	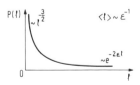
II	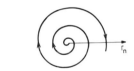 $$r_{n+1} = (1 + \varepsilon)r_n + u r_n^3$$ $$\theta_{n+1} = \theta_n + \Omega$$	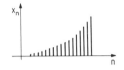 spirals	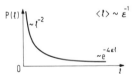
III	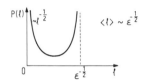 $$x_{n+1} = -(1 + \varepsilon)x_n - u x_n^3 \quad \text{alternates}$$		

Below we present a derivation of $P(l)$ and describe two representative experiments in which type-I intermittency has been detected. Type-II intermittency has (to the best of our knowledge) not yet been found in a real experiment. This section closes with brief report on the first experimental observation of type-III intermittency.

Distribution of Laminar Lengths

We assume that the signal is randomly reinjected (with a probabilitiy $\hat{P}(x_0)$) into the laminar régime in such a way that we can use eq. (4.54):

$$P(l) = \hat{P}(x_0) \left| \frac{dx_0}{dl} \right| . \qquad (4.66)$$

In order to obtain $x_0(l)$, we approximate, as in (4.9), the Poincaré map for type-I intermittency (see Table 4)

$$x_{n+1} = \varepsilon + x_n + u x_n^2 \qquad (4.67)$$

in the laminar region by the differential equation

$$\frac{dx}{dl} = \varepsilon + u x^2 . \qquad (4.68)$$

This yields by integration

$$l = \frac{1}{\sqrt{\varepsilon u}} \left[\arctan \left[\frac{c}{\sqrt{\varepsilon/u}} \right] - \arctan \left[\frac{x_0}{\sqrt{\varepsilon/u}} \right] \right] \qquad (4.69)$$

where c is the maximum value of $x(l)$ in the laminar régime (see Fig. 58). $P(l)$ follows from eqns. (4.66) and (4.69):

$$P(l) = \frac{\varepsilon}{2c} \left\{ 1 + \tan^2 \left[\arctan \left[\frac{c}{\sqrt{\varepsilon/u}} \right] \right] - l\sqrt{\varepsilon u} \right\} \qquad (4.70)$$

and

$$\langle l \rangle = \int_0^\infty dl\, P(l)\, l \sim \varepsilon^{-1/2} \quad \text{for} \quad \varepsilon \to 0 . \qquad (4.71)$$

The distributions $P(l)$ for the two other types of intermittency are obtained in a similar way, with the net results

$$P(l) \sim \frac{\varepsilon^2 e^{4\varepsilon l}}{(e^{4\varepsilon l} - 1)^2} \qquad \text{for type II} \qquad (4.72)$$

and

$$P(l) \sim \frac{\varepsilon^{3/2} e^{4\varepsilon l}}{(e^{4\varepsilon l} - 1)^{3/2}} \qquad \text{for type III.} \qquad (4.73)$$

For type-II intermittency eq. (4.66) has to be replaced by $P(l) = \hat{P}(r_0) r_0 | \, dr_0/dl |$ because the Poincaré map is two-dimensional.

Type-I Intermittency

Fig. 63 shows the vertical velocity as a function of time for a Bénard experiment. The signal shows a behavior which is typical for type-I intermittency.

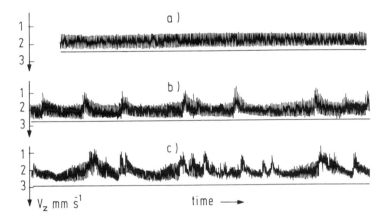

Fig. 63: Intermittency for a Bénard experiment: The vertical velocity component measured in the middle of a Bénard cell changes with increasing Rayleigh number from periodic motion (a) via intermittent motion (b) to chaos (c) (after Bergé et al., 1980).

The nonlinear RCL-oscillator described on page 75 also displays the intermittency route. Type-I intermittency is indicated in Fig. 64 by the Poincaré map, the scaling behavior of the lengths of the laminar regions, and the maximum in $P(l)$ for $l > 0$.

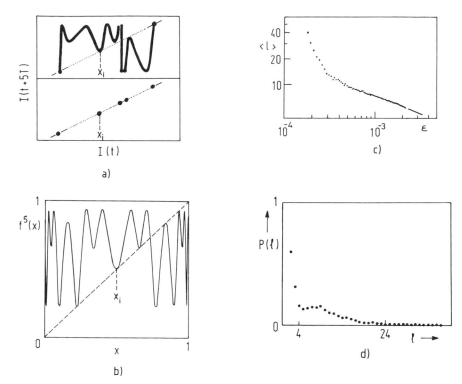

Fig. 64: Intermittency in the nonlinear *RCL*-oscillator: a)$I(t + 5T)$ versus $I(t)$ which corresponds to the fifth iterate of the logistic map at tangency which is shown in b). c) The measured averaged length for which the laminar regions scales like $\langle l \rangle \propto \varepsilon^{-0.43}$ (where $\varepsilon \sim V_0 - V_c$) is in reasonable agreement with the prediction of Manneville and Pomeau $\langle l \rangle \propto \varepsilon^{-0.5}$. d) $P(l)$ vs. laminar lengths l (in units of $5T$) for $\varepsilon = 2.5 \cdot 10^{-4}$. (After Jeffries and Pérez, 1982.)

Type-III Intermittency

Type-III intermittency has first been observed by M. Dubois, M. A. Rubio and P. Bergé (1983) in Bénard convection in a small rectangular cell. They measured the local horizontal temperature gradient via the modulation of a light beam that was sent through the cell.

Fig. 65 as shows the time dependence of the light intensity that is characteristic for type-III intermittency. The intermittency appears simultaneously with a period-doubling bifurcation. One observes the growth of a subharmonic amplitude together with a decrease of the fundamental amplitude. When the subharmonic amplitude reaches a high value, the signal looses its regularity, and turbulent bursts appear.

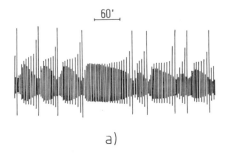

60'

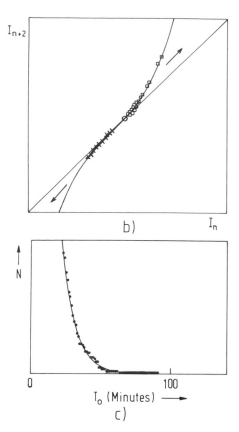

I_{n+2}

b) I_n

Fig. 65: a) Time dependence of the light intensity which is roughly proportional to the local horizontal temperature gradient. b) Poincaré map I_{n+2} versus I_n constructed from the data in a) for $\varepsilon = 0.098$. The amplitudes of the light modulation in the turbulent bursts have not been drawn. Note that the "ghost of the fixed point" o is purely repulsive.
c) Number N of laminar lengths with $l > T_0$

i.e. $N = \int_{T_0}^{\infty} P(l)\,dl$ versus T_0. The experimental points agree with the line obtained from (4.73) for $\varepsilon = 0.098$. (After Dubois et al., 1983).

N

0   100

T_0 (Minutes) ⟶

c)

By plotting subsequent maxima I_n of both the subharmonic mode (even n, crosses) and the fundamental mode (odd n, squares), the Poincaré map shown in Fig. 65 b is obtained. Its form can be described by

$$I_{n+2} = (1 + 2\varepsilon)I_n + bI_n^3 \tag{4.74}$$

where b is a constant and $\varepsilon \propto (R - R_c)$ measures the distance to the critical Rayleigh number R_c (which corresponds to the threshold of the intermittent behavior). Equation (4.74) can be derived from the map

$$I_{n+1} = f(I_n) \equiv -(1 + \varepsilon)I_n - uI_n^3 \tag{4.75}$$

with $b = u(2 + 4\varepsilon)$. Its eigenvalue

$$\lambda = f'(0) = -(1 + \varepsilon) \tag{4.76}$$

crosses the unit circle at -1, which again signals type-III intermittency according to Table 5.

5 Strange Attractors in Dissipative Dynamical Systems

In the first part of this chapter we show that nonlinear dissipative dynamical systems lead naturally to the concept of a strange attractor. In Section 2, the Kolmogorov entropy is introduced as the fundamental measure for chaotic motion. Section 3 deals with the problem of how much information about a strange attractor can be obtained from a measured random signal. We discuss the reconstruction of the trajectory in phase space from the measured time series of a single variable and introduce generalized dimensions and entropies. It is demonstrated how this quantities can be obtained from a measurement and how one can extract from them the distribution of singularities in the invariant measure that characterizes the static structure of a strange attractor and the fluctuation spectrum of the Kolmogorov entropy which describes the dynamical evolution of the trajectory on the attractor. Finally we present in the last chapter a collection of pictures of strange attractors and fractal boundaries.

5.1 Introduction and Definition of Strange Attractors

In this section, we consider dissipative systems that can be described either by flows or maps. Let us begin with dissipative flows. These are described by a set of autonomous first-order differential equations,

$$\dot{\vec{x}} = \vec{F}(\vec{x}), \qquad \vec{x} = (x_1, x_2, \ldots x_d) \tag{5.1}$$

and the term dissipative means that an arbitrary volume element V enclosed by some surface S in phase space $\{\vec{x}\}$ contracts. The surface S evolves by having each point on it follow an orbit generated by (5.1). This yields, by the divergence theorem,

$$\frac{dV}{dt} = \int_V d^d x \left(\sum_{i=1}^d \frac{\partial F_i}{\partial x_i} \right) \tag{5.2}$$

and dissipative systems are defined by $dV/dt < 0$.

An example of this kind of flow is given by the Lorenz model

$$
\begin{aligned}
\dot{X} &= -\sigma X + \sigma Y \\
\dot{Y} &= -XZ + rX - Y \\
\dot{Z} &= XY - bZ
\end{aligned}
\tag{5.3}
$$

for which one finds via (5.2)

$$
\frac{dV}{dt} = -(\sigma + 1 + b)V < 0; \quad (\sigma > 0, b > 0)
\tag{5.4}
$$

i. e. the volume element contracts exponentially in time:

$$
V(t) = V(0)\,e^{-(\sigma + 1 + b)t}.
\tag{5.5}
$$

If, on the other hand, the trajectory generated by the equations of the Lorenz model for $r = 28, \sigma = 10, b = 8/3$ is considered (see Fig. 66), one finds that it is a) attracted to a bounded region in phase space; b) the motion is erratic; i. e., the trajectory makes one loop to the right, then a few loops to the left, then to the right, etc.; and c) there is a sensitive dependence of the trajectory on the initial conditions; i. e., if instead of (0, 0.01, 0) an adjacent initial condition is taken, the new solution soon deviates from the old, and the number of loops is different. Fig. 67 shows a plot of the nth maximum M_n of Z versus M_{n+1}. The resulting map is approximately triangular, which corresponds, according to the material discussed in Chapter 2, to a chaotic sequence of M_n's.

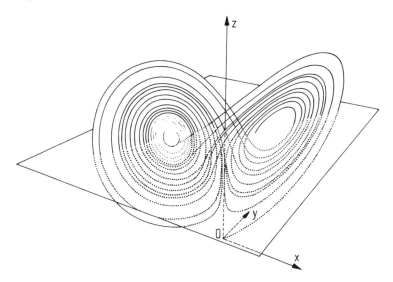

Fig. 66: The Lorenz attractor, after a computer calculation by Lanford (1977).

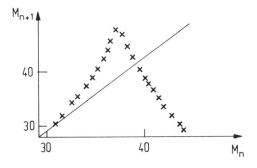

Fig. 67: Successive maxima of Z of the Lorenz attractor (after E. N. Lorenz, 1963).

Summarizing: The trajectory depends sensitively on the initial conditions; it is chaotic; it is attracted to a bounded region in phase space; and (according to eq. (5.4)) the volume of this region contracts to zero. This means that the flow of the three-dimensional Lorenz system generates a set of points whose dimension is less than three; i. e., its volume in three-dimensional space is zero. At first sight, one might think of the next lower integer dimension, two. However, this is forbidden by the *Poincaré-Bendixson* theorem which states that there is no chaotic flow in a bounded region in two-dimensional space. We refer, e. g., to the monograph by Hirsch and Smale (1965) for a rigorous proof of this theorem. However, Fig. 68 makes it plausible that both the continuity of the flow lines and the fact that a line divides a plane into two parts restrict the trajectories in two dimensions so strongly that the only possible attractors for a bounded region are limit cycles or fixed points. The solution to this problem is that the set of points to which the trajectory in the Lorenz system is attracted, the so-called Lorenz attractor, has a Hausdorff dimension which is noninteger and lies between two and three (the precise value is $D = 2.06$). This leads, in a natural way, to the concept of a strange attractor which appears in a large variety of physical, nonlinear systems.

Fig. 68: Self-trapping of a flow line in a bounded region of the plane. Exponential separation of points is at variance with continuity (note the opposing arrows).

A *strange attractor* has the following properties (a more formal definition can be found in the review articles by Eckmann and Ruelle, 1985):

a) It is an attractor, i. e., a bounded region of phase space $\{\vec{x}\}$ to which all sufficiently close trajectories from the so-called basin of attraction are attracted asymptotically for long enough times. We note that the basin of attraction can have a very complicated structure (see the pictures in Sect. 5.4). Furthermore, the attractor itself should be indecomposable; i. e., the trajectory should visit every point on the attractor in the course of time. A collection of isolated fixed points is no single attractor.

b) The property which makes the attractor strange is the sensitive dependence on the initial conditions; i.e., despite the contraction in volume, lengths need not shrink in all directions, and *points, which are arbitraily close initially, become exponentially separated at the attractor for sufficiently long times.* This leads to a positive Kolmogorov entropy, as we shall see in the next section.

All strange attractors that have been found up to now in dissipative systems have fractal Hausdorff dimensions. Since there exists no generally accepted formal definition of a strange attractor (Ruelle, 1980; Mandelbrot, 1982), it is not yet clear whether a fractal Hausdorff dimension follows already from a)–b) or should be additionally required for a strange attractor.

A strange attractor arises typically when the flow contracts the volume element in some directions, but stretches it along the others. To remain confined to a bounded domain, the volume element is folded at the same time. By analogy to the broken linear maps in Chapter 2, this stretching and backfolding process produces a chaotic motion of the trajectory at the strange attractor (see also Fig. 69).

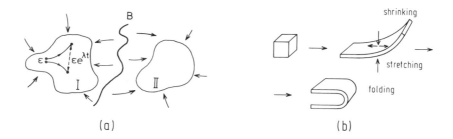

(a)   (b)

Fig. 69: a) Two strange attractors I and II with different basins of attraction separated by a boundary B. b) Deformation of a volume element on a strange attractor with increasing time. This leads to the foliated fractal structure shown in Fig. 90 c.

Because the definition given above describes the properties of a set of points, the concept of a strange attractor is not confined to flows, and dissipative maps can also generate strange attractors. A map

$$\vec{x}(n+1) = \vec{G}[\vec{x}(n)] ; \quad \vec{x}_1(n) = [x_1(n), \dots x_d(n)] \qquad (5.6a)$$

is called dissipative if it leads to a contraction of volume in phase space; i.e., if the absolute value of its Jacobian J, by which a volume element is multiplied after each iteration, is smaller than unity:

$$|J| = \left| \det\left(\frac{\partial G_i}{\partial x_j} \right) \right| < 1 . \qquad (5.6b)$$

The Poincaré-Bendixson theorem that restricts the dimension of strange attractors generated by flows to values larger than two does not hold for maps. This is because maps generate discrete points and the restrictions imposed by the continuity of the flow are lifted. Dissipative maps can therefore lead to strange attractors that also have dimensions smaller than two.

Let us consider two illustrative examples which, because of their lower dimensionality, are easier to visualize than the Lorenz attractor.

Baker's Transformation

Fig. 70 shows the usual baker's transformation, which is an area preserving map (reminiscent of a baker kneading dough), and the non-area preserving, dissipative baker's transformation. The mathematical expression for the latter is

$$x_{n+1} = 2x_n \bmod 1 \qquad (5.7a)$$

$$y_{n+1} = \begin{cases} a y_n & \text{for} \quad 0 \le x_n < \dfrac{1}{2} \\[2mm] \dfrac{1}{2} + a y_n & \text{for} \quad \dfrac{1}{2} \le x_n \le 1 \end{cases} \qquad (5.7b)$$

where $a < 1/2$.

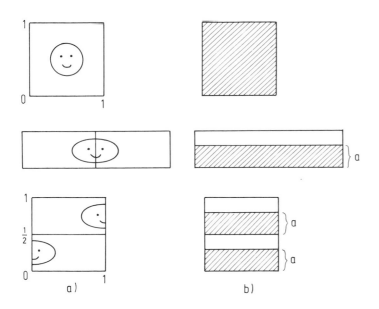

Fig. 70: a) Baker's transformation; b) dissipative baker's transformation.

The first equation (5.7a) is our old friend from Chapter 2: the transformation σ which leads to the Bernoulli shift. It has a Liapunov exponent (in x-direction), $\lambda_x = \log 2 > 0$, which leads to the sensitive dependence on the initial conditions, and makes the object resulting from repeated applications of this map to the unit square a strange attractor. The attractor is an infinite sequence of horizontal lines, and its basin of attraction consists of all points within the unit square. The Liapunov exponent in the y-direction is $\lambda_y = \log a < 0$, and lengths are contracted in this direction such that the net result (of the stretching in x- and shrinking in y-direction) is a volume contraction, as required for a dissipative map.

The Hausdorff dimension D_B of this strange attractor can be calculated as follows: In the x-direction the attractor is simply one-dimensional (as the map $\sigma(x)$ of Chapter 2). The Hausdorff dimension in the y-direction follows from its definition

$$\lim_{l \to 0} N(l) \propto l^{-D_y} \tag{5.8}$$

and from the self-similarity of the attractor in the vertical direction, shown in Fig. 70b. This yields

$$\frac{N(a)}{N(a^2)} = \frac{1}{2} = a^{-D_y} \to D_y = \log\left(\frac{1}{2}\right)\bigg/\log a \tag{5.9}$$

and finally

$$D_B = 1 + D_y = 1 + \frac{\log 2}{|\log a|} \, . \tag{5.10}$$

Dissipative Hénon Map

This is the two-dimensional analogue of the logistic map introduced by Hénon (1976), and we recall its recursion relation from Capter 1

$$x_{n+1} = 1 - a x_n^2 + y_n \tag{5.11a}$$

$$y_{n+1} = b x_n \, . \tag{5.11b}$$

This map is area contracting, i.e., is dissipative for $|b| < 1$ because its Jacobian is just

$$\left| \det \begin{pmatrix} -2 a x_n & 1 \\ b & 0 \end{pmatrix} \right| = |b| \, . \tag{5.12}$$

The action of the map is shown in Fig. 71.

Let us now examine its iterates for, e. g., $b = 0.3$, $a = 1.4$. Fig. 72 a shows the result of an iteration with 10^4 steps, and we have indicated the dynamics by enumerating some successive points on the attractor that looks like a very tangled curve. Figs. 72 b–c show details of the regions inside the box of the previous figure and reveal the selfsimilar structure of the attractor. The Hausdorff dimension of the Hénon attractor is: $D (a = 1.4, b = 0.3) = 1.26$. This result was obtained by placing a square net of width l over the diagram, counting the number $N(l)$ of squares occupied by points, and forming $D = - \lim_{l \to 0} \log N(l)/\log l$. If Fig. 72c is resolved into six „leaves“, then

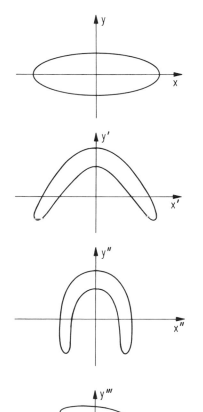

Initial ellipse

Area preserving bending

$T_1 : x' = x$
$\quad\quad y' = 1 - ax^2 + y$

Contraction in the x-direction

$T_2 : x'' = bx'$
$\quad\quad y'' = y'$

Rotation by $90°$

$T_3 : x''' = \quad y''$
$\quad\quad y''' = -x''$

Fig. 71: Decomposition of the action of the Hénon map $T = T_3 \cdot T_2 \cdot T_1$ on an ellipse.

the relative probability of each leaf can be estimated by simply counting its number of points. The height of each bar in Fig. 72 d is the relative probability, and the width is the thickness of the corresponding leaf.

The different heights of the bars in Fig. 72 d show that the *Hénon attractor is in-homogeneous.* This inhomogeneity cannot be described by the Hausdorff dimension alone and in the following we shall therefore introduce an infinite set of dimensions which characterize the static structure (i.e. the distribution of points) of the attractor. However, before this step, it is useful to discuss the Kolmogorov entropy that describes the dynamical behavior at the strange attractor.

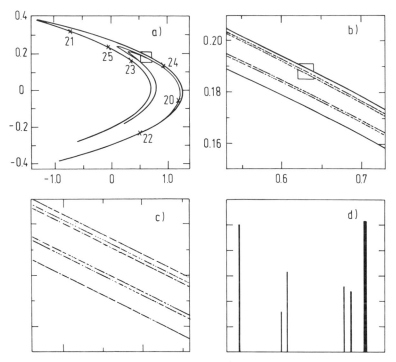

Fig. 72: a) The Hénon attractor for 10^4 iterations. Some successive iterates have been numbered to illustrate their erratic movement on the attractor. b), c) Enlargements of the squares in the preceding figure. d) The height of each bar is the relative probability to find a point in one of the six leaves in c). (After Farmer, 1982 a, b.)

5.2 The Kolmogorov Entropy

The Kolmogorov entropy (Kolmogorov, 1959) is the most important measure by which chaotic motion in (an arbitrary-dimensional) phase space can be characterized.

Before we introduce this quantity, it is useful to recall that the thermodynamic entropy S measures the disorder in a given system. A simple example, for a system where

5 Strange Attractors in Dissipative Dynamical Systems

S increases, is that of gas molecules that are initially confined to one half of a box, but are then suddenly allowed to fill the whole container. The disorder in this system increases because the molecules are no longer separated from the other half of the box. This increase of disorder is coupled with an increase of our ignorance about the state of the system (before the confinement was lifted, we knew more about the positions of the molecules).

More precisely, the entropy S, which can be expressed as

$$S \propto - \sum_i P_i \log P_i \tag{5.13}$$

where $\{P_i\}$ are the probabilities of finding the system in states $\{i\}$, measures, according to Shannon et al. (1949) (see Appendix F), the information needed to locate the system in a certain state i^*; i.e., S is a measure of our ignorance about the system.

This example from statistical mechanics shows that disorder is essentially a concept from information theory. It is therefore not too surprising that the Kolmogorov entropy K, which measures "how chaotic a dynamical system is", can also be defined by Shannon's formula in such a way that K becomes proportional to the rate at which information about the state of the dynamical system is lost in the course of time.

Definition of K

K can be calculated as follows (Farmer, 1982a, b): Consider the trajectory $\vec{x}(t) = [x_1(t), \ldots x_d(t)]$ of a dynamical system on a strange attractor and suppose that the d-dimensional phase space is partitioned into boxes of size l^d. The state of the system is now measured at intervals of time τ. Let $P_{i_0 \ldots i_n}$ be the joint probability that $\vec{x}(t = 0)$ is in box i_0, $\vec{x}(r = \tau)$ that it is in box i_1, $\ldots$, and $\vec{x}(t + n\tau)$ that it is in box i_n. According to Shannon, the quantity

$$K_n = - \sum_{i_0 \ldots i_n} P_{i_0 \ldots i_n} \log P_{i_0 \ldots i_n} \tag{5.14}$$

is proportional to the information needed to locate the system on a special trajectory $i_0^* \ldots i_n^*$ with precision l (if one knows a priori only the probabilities $P_{i_0 \ldots i_n}$). Therefore, $K_{n+1} - K_n$ is the additional information needed to predict in which cell i_{n+1}^* the system will be if we know that it was previously in $i_0^* \ldots i_n^*$. This means, that $K_{n+1} - K_n$ measures our loss of information about the system from time n to time $n + 1$.

The K-entropy is defined as the average rate of loss of information:

$$K = \lim_{\tau \to 0} \lim_{l \to 0} \lim_{N \to \infty} \frac{1}{N\tau} \sum_{n=0}^{N-1} (K_{n+1} - K_n) =$$

$$= - \lim_{\tau \to 0} \lim_{l \to 0} \lim_{N \to \infty} \frac{1}{N\tau} \sum_{i_0 \ldots i_{N-1}} P_{i_0 \ldots i_{N-1}} \log P_{i_0 \ldots i_{N-1}} . \tag{5.15}$$

The limit $l \to 0$ (which has to be taken *after* $N \to \infty$) makes K independent of the particular partition. For maps with discrete time steps $\tau = 1$, the limit $\tau \to 0$ is omitted.

Table 8 shows that K is indeed a useful measure of chaos. K becomes zero for regular motion, it is infinite in random systems, but it is a constant larger than zero if the system displays deterministic chaos.

Table 8: K-entropies for (one-dimensional) regular, chaotic and random motion.

Regular motion

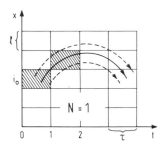

Initially adjacent points stay adjacent

$$P_{i_0} = l, \quad P_{i_0 i_1} = l \cdot 1$$
$$K = 0$$

Chaotic motion

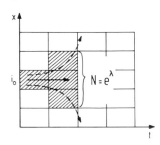

Initially adjacent points become exponentially separated

$$P_{i_0} = l, \quad P_{i_0 i_1} = l \, e^{-\lambda}$$
$$K = \lambda > 0$$

Random motion

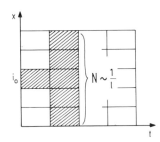

Initially adjacent points are distributed with equal probability over all newly allowed intervals

$$P_{i_0} = l, \quad P_{i_0 i_1} \propto l^2$$
$$K \propto -\log l \to \infty$$

Here we assumed for simplicity that a) $P_{i_0 i_1}$ factorizes into $P_{i_0} \cdot (1/N)$ where N is the number of possible new intervals which evolve from i_0 and b) $K_{n+1} - K_n = K_1 - K_0$ for all n.

Connection of K to the Liapunov Exponents

For one-dimensional maps, K is just the positive Liapunov exponent (see Table 8 and eq. (2.12)). In higher dimensional systems, we loose information about the system because the cell in which it was previously located spreads over new cells in phase space at a rate which is determined by the positive Liapunov exponents (see Fig. 73). It is therefore plausible that the rate K at which information about the system is lost is equal to the (averaged) sum of positive Liapunov exponents (Pesin, 1977):

$$K = \int d^d x \rho(\vec{x}) \sum_i \lambda_i^+(\vec{x}) . \qquad (5.16)$$

Here $\rho(\vec{x})$ is the invariant density of the attractor. In most cases, the λ's are independent of $\vec{x}$; the integral then becomes unity, and K reduces to a simple sum.

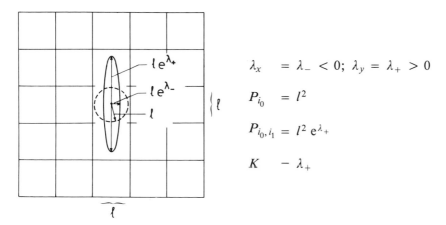

$$\lambda_x = \lambda_- < 0; \ \lambda_y = \lambda_+ > 0$$

$$P_{i_0} = l^2$$

$$P_{i_0, i_1} = l^2 e^{\lambda_+}$$

$$K - \lambda_+$$

Fig. 73: A two-dimensional map transforms a small circle into an ellipse with minor and major radii distorted according to the Liapunov exponents λ_x and λ_y. Note that e^{λ_-} does not enter K because, due to this exponent, *no* new cells are covered after one time step.

The definition of the Liapunov exponent λ for a one-dimensional map $G(x)$ (see eq. (2.9)),

$$e^\lambda = \lim_{N \to \infty} \left(\prod_{n=0}^{N-1} \left| \frac{dG}{dx_n} \right| \right)^{1/N} \qquad (5.17)$$

can be easily generalized to d dimensions, where we have d exponents for the different spatial directions,

$$(e^{\lambda_1}, e^{\lambda_2} \dots e^{\lambda_d}) = \lim_{N \to \infty} (\text{magnitude of the eigenvalues of } \prod_{n=0}^{N-1} J(\vec{x}_n))^{1/N} \quad (5.18)$$

and

$$J(\vec{x}) = \left(\frac{\partial G_i}{\partial x_j} \right) \tag{5.19}$$

is the Jacobian matrix of the map $\vec{x}_{n+1} = \vec{G}(\vec{x}_n)$.

Note that the eigenvalues $\{\lambda_i\}$ of the Jacobian matrix are invariant under coordinate transformations in phase space, i.e. from (5.16), K is also invariant, as one would expect for such an important physical quantity.

Let us briefly comment on the computation of Liapunov exponents for flows. First, there is a difference in the calculation of Liapunov exponents for maps (λ_M) and flows (λ_F) which can be explained by the following trivial example. The Liapunov exponent λ_M of the map

$$x_{n+1} = a x_n \rightarrow x_n = e^{n \log a} x_0 \tag{5.20}$$

is obviously $\lambda_M = \log a$. Whereas one obtains for the flow

$$\dot{x} = ax \rightarrow x(t) = e^{at} x(0) , \tag{5.21}$$

the result that nearby trajectories separate with rate a i.e. the Liapunov exponent λ_F is simply $\lambda_F = a$. (Both examples show no chaos, of course, because backfolding is missing).

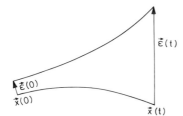

Fig. 74: Exponential separation of nearby trajectories in phase space (schematically).

For general flows described by an autonomous differential equation

$$\dot{\vec{x}} = \vec{f}(\vec{x}) , \tag{5.22}$$

the difference $\vec{\varepsilon}(t)$ of infinitesimal neighbored trajectories (see Fig. 74) develops according to

$$\dot{\vec{\varepsilon}} = M(t) \vec{\varepsilon} \tag{5.23}$$

where

$$M_{ij}(t) = \frac{\partial f_i}{\partial x_j} \{\vec{x}[t, \vec{x}(0)]\} \tag{5.24}$$

is the Jacobian matrix taken at the point $\vec{x}(t)$. Therefore, in order to integrate eq. (5.23) one has to integrate eq. (5.22) first to know $\vec{x}[t, \vec{x}(0)]$. However, eq. (5.23) can be integrated formally yielding

$$\vec{\varepsilon}(t) = \left\{ \hat{T} \exp \left[\int_0^t dt' M(t') \right] \right\} \vec{\varepsilon}(0) \equiv L(t)\vec{\varepsilon}(0) \tag{5.25}$$

where the time ordering operator $\hat{T}$ has to be introduced because the matrices $M(t)$ and $M(t')$ usually do not commute at different times t and t'. The Liapunov exponents $\lambda_1 \ldots \lambda_d$ of the flow are, in analogy to eq. (5.18), defined as

$$(e^{\lambda_1}, e^{\lambda_2}, \ldots e^{\lambda_d}) = \lim_{t \to \infty} (\text{magnitude of the eigenvalues of } L(t))^{\frac{1}{t}}. \tag{5.26}$$

The Liapunov exponents in eq. (5.26) generally depend on the choice of the intial point $\vec{x}(0)$. Even if $\vec{x}(t)$ moves on a strange attractor, a change in $\vec{x}(0)$ could place the system into the basin of attraction of another attractor with a different set of λ_i's (see e.g. Fig. 69).

We will not discuss all numerical methods which have been developed in order to extract the Liapunov exponents from eqns. (5.22–24) (see the References of this section for some examples), but only explain the simplest method which yields the largest Liapunov exponent λ_m.

Expanding in eq. (5.25), $\vec{\varepsilon}(0)$ with respect to the eigenvectors $\vec{e}_j$ of $L(t)$ i.e.

$$\vec{\varepsilon}(0) = \sum_{j=1}^d a_j \vec{e}_j; \quad a_j = \vec{e}_j \cdot \vec{\varepsilon}(0) \tag{5.27}$$

we obtain by using

$$L(t)\vec{e}_j \propto e^{\lambda_j t} \vec{e}_j \quad \text{for} \quad t \to \infty \tag{5.28}$$

via eq. (5.25):

$$|\vec{\varepsilon}(t)| = \left| \sum_{j=1}^d a_j \vec{e}_j e^{\lambda_j t} e^{i\psi_j t} \right| \propto e^{\lambda_m t} \quad \text{for} \quad t \to \infty. \tag{5.29}$$

Here, ψ_j denotes the phase angle of the j'th eigenvalue of $L(t)$, which can be complex, and $e^{\lambda_m t}$ dominates the sum in eq. (5.29) because the remaining terms decay as $\exp[-|\lambda_m - \lambda_j| t]$. In order to obtain λ_m, one could therefore start with any randomly chosen value for $\vec{\varepsilon}(0)$, calculate $\vec{\varepsilon}(t)$ by numerical integration of eqns. (5.22–24), and extract λ_m via eq. (5.29). To avoid overflow in the computer, this is usually done in steps as shown in Fig. 75.

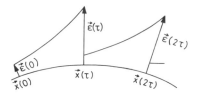

Fig. 75: Numerical calculation of the largest Liapunov exponent λ_m. To avoid overflow, one calculates the divergence of nearby trajectories for finite timesteps τ renormalizes $\vec{\varepsilon}(i\tau)$ to unity after each step and takes the average:

$$\varepsilon(\tau) = \vec{\varepsilon}(0)\exp(\lambda_1\tau) ; \quad \vec{\varepsilon}(2\tau) = [\vec{\varepsilon}(\tau)/|\vec{\varepsilon}(\tau)|]e^{\lambda_2\tau}\ldots$$

$$\lambda_m = \lim_{n\to\infty}\frac{1}{n}\sum_{i=1}^{n}\lambda_i = \lim_{n\to\infty}\frac{1}{n\tau}\sum_{i=1}^{n}\log|\vec{\varepsilon}(\tau i)| \ .$$

Plate XVII, at the beginning of this book, and Fig. 76 display the parameter dependence of λ_m for the driven pendulum with an additional torque and for the Lorenz model, respectively. In both cases, one observes a sensitive dependence of order $\lambda_m < 0$ and chaos $\lambda_m > 0$ on the parameter values.

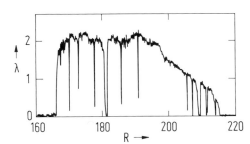

Fig. 76: The largest Liapunov exponent of the Lorenz model (eqns. (1.4) for $\sigma = 10$, $b = 8/3$) as a function of the parameter $r = R$ (after K. Schmidt priv. comm.).

Average Time over which the State of a Chaotic System can be Predicted

The K-entropy also determines the average time over which the state of a system, displaying deterministic chaos, can be predicted. Consider, e. g., the simple one-dimensional triangular map in Fig. 15b, which is confined to the unit square. After n time steps, an interval l increases to $L = le^{\lambda n}$. If L becomes larger than 1, we can no longer locate the trajectory in [0, 1], and all we can say is that the system has a probability

$$\rho_0(x)\,\mathrm{d}x \tag{5.30}$$

of being in an interval $[x, x + \mathrm{d}x] \in [0, 1]$, where $\rho_0(x)$ is the invariant density of the system. In other words, precise predictions about the state of this system are only possible for times n that are smaller T_m:

$$l \mathrm{e}^{\lambda T_m} = 1 \rightarrow T_m = \frac{1}{\lambda} \log \left(\frac{1}{l}\right). \tag{5.31}$$

Above T_m, one can only make statistical predictions. Eq. (5.31) can be generalized to higher dimensional dynamical systems by replacing λ by the K-entropy (Farmer, 1982 a):

$$T_m \propto \frac{1}{K} \log \left(\frac{1}{l}\right). \tag{5.32}$$

Note that the precision l, with which the initial state is located, only influences T_m logarithmically. Let us summerize our results about the K-entropy:

- It measures the average rate at which information about the state of a dynamical system is lost with time.

- For one-dimensional maps, it is equal to the Liapunov exponent. In higher dimensional systems, K measures the average deformation of a cell in phase space and becomes equal to the integral over phase space of the sum of the positive Liapunov exponents.

- It is inversely proportional to the time interval over which the state of a chaotic system can be predicted.

Furthermore, in the next section, we shall show that K can be directly obtained by measuring the time dependence of one component of a chaotic system. These results show that the K-entropy is *the* fundamental quantity by which chaotic motion can be characterized, and we define a strange attractor as an attractor with a positive K-entropy.

5.3   Characterization of the Attractor by a Measured Signal

Having experimentally observed a seemingly chaotic signal, one wants to know what information it contains about the strange attractor. To provide an answer, we proceed in several steps.

First, we will explain the result of Takens (1981) who has shown that, after the transients have died out, one can reconstruct the trajectory on the attractor (the whole time

---

dependent vector $\vec{x}(t) = [x_1(t), x_2(t) \ldots]$ in phase space) from the measurement of a single component, say $x_1(t)$. A knowledge of the time series of one variable is therefore sufficient to reconstruct the statical and dynamical properties of the strange attractor.

Since the whole trajectory contains too much information, we then follow a series of papers by Grassberger, Hentschel and Procaccia (1983), Halsey et al. (1986), Eckmann and Procaccia (1986) and introduce a set of averaged coordinate invariant numbers (generalized dimensions, entropies, and scaling indices) by which different strange attractors can be distinguished. For this purpose, we divide the attractor into boxes of linear dimension l, and denote by p_i the probability that the trajectory on the strange attractor visits box i. By averaging powers of the $p_i's$ over all boxes, we obtain the generalized dimensions D_q, defined by

$$D_q = - \lim_{l \to 0} \frac{1}{q-1} \left| \frac{1}{\log l} \right| \log \left(\sum_i p_i^q \right) \tag{5.33}$$

that are formally similar to the free energy F_β of ordinary equilibrium thermodynamics:

$$F_\beta = - \lim_{N \to \infty} \frac{1}{\beta} \frac{1}{N} \log \left[\sum_i (e^{-E_i})^\beta \right] \tag{5.34}$$

where E_i are the energy levels of the system, N is its particle number, and β is the inverse temperature. Since $\sum_i p_i^q$, which appears in eq. (5.33), is for $q > 1$ the total probability that q points of the attractor are within one box, it is obvious that the D_q's measure correlations between different points on the attractor and are therefore useful in characterizing its inhomogeneous static structure.

But, it will be shown below that the (negative) Legendre transform $f(\alpha)$ of the D_q's (more precisely of $(q-1)D_q$):

$$f(\alpha) = - (q-1) D_q + q\alpha \tag{5.35a}$$

$$\alpha = \frac{\partial}{\partial q} [(q-1)D_q] \tag{5.35b}$$

is more appropriate to describe universal properties of strange point sets.

Let us briefly explain the meaning of $f(\alpha)$ (its connection to the $D_q's$ will be shown below). Assuming ergodicity, the probabilities p_i are, by construction, related to the invariant density $\rho(\vec{x})$ of the attractor:

$$p_i = \int_{|\vec{x}_i - \vec{x}| \le l} d^d x \rho(\vec{x}) \tag{5.36}$$

where $\vec{x}_i$ denotes the center of box i. If $p_i(l)$ diverges for $l \to 0$ as

$$p_i(l \to 0) \propto l^{\alpha_i} \tag{5.37}$$

the invariant density has according to eq. (5.36) at $\vec{x}_i$ a singularity whose strength is characterized by α_i. Since different points $\vec{x}_i$ on the attractor can have different strengths α_i, it is useful to introduce a function $f(\alpha)$ which measures the Hausdorff dimension of the set of points $\{\vec{x}_i\}$ on the attractor which have the same strength of singularity α. $f(\alpha)$ characterizes the static distribution of points on the attractor and can therefore also be used for point sets which are not generated dynamically (see Fig. 77).

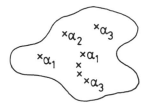

Fig. 77: The invariant measure has on a strange attractor different power law singularities. $f(\alpha)$ measures the Hausdorff dimension of the set of points with the same power α.

In order to describe the dynamical behavior of the trajectory on the attractor, we use the quantities $P_{i_0 \ldots i_n}$ which we introduced already to define the K-entropy via eq. (5.14). The $P_{i_0 \ldots i_n}$'s measure the probability that the trajectory visits a certain sequence $i_0 \ldots i_n$ of boxes of size l in time n. By playing with these variables the same game as with the P_i's, we introduce generalized entropies K_q via:

$$K_q = - \lim_{l \to 0} \lim_{n \to \infty} \frac{1}{q-1} \frac{1}{n} \log \sum_{i_0 \ldots i_n} P_{i_0 \ldots i_n}^q \tag{5.38}$$

and show that their Legendre transform $g(\lambda)$ is connected to the fluctuations of the K-entropy around its mean value K_1 given by eq. (5.15).

It will also be demonstrated that both quantities, the D_q's and the K_q's (and therefore $f(\alpha)$ and $g(\lambda)$) can be extracted from a time series of a single variable. Two further important quantities, which can be obtained in this way, are the embedding dimension of the attractor, that is, the dimension of the space with the lowest integer dimension, which contains the attractor, and the amplitude of white noise on the signal. Thus irregularities originating from deterministic motion on the attractor can be separated from disturbing white noise.

Reconstruction of the Attractor from a Time Series

It is not always possible to measure all components of the vector $\vec{x}(n)$ simultaneously. This clearly holds for an infinite-dimensional system. If we define the dimension of a system by the number of initial conditions, then the so-called Mackey-Glass equation (Mackey and Glass, 1977)

$$\dot{x} = \frac{ax(t - \tau)}{1 + \{x(t - \tau)\}^{10}} - bx(t) \tag{5.39}$$

(which describes the regeneration of blood cells) obviously provides a simple example of an infinite-dimensional system, because all the $x(t)$-values in the interval $t, t - \tau$ have to be known (as initial conditions) to solve it. How do we proceed in this, or the less difficult case, where we have an attractor embedded in d-dimensional space, but measure only one component of the signal?

It has been shown by Takens (1981) that one can *reconstruct certain properties of the attractor* in phase space *from the time series of a single component*. Instead of the rather cumbersome proof, we present the following simplified argument. As an example, consider a two-dimensional flow generated by

$$\frac{d}{dt} \vec{x} = \vec{F}(\vec{x}) \qquad \vec{x} = \{x, y\} \tag{5.40}.$$

Every point $\{x(t + \tau), y(t + \tau)\}$ then originates uniquely from a point $\{x(t), y(t)\}$, and the relation between both points is one-to-one because the trajectories do not cross (otherwise the trajectory would not be determined uniquely by the initial conditions). Next, we construct a sequence of vectors

$$\vec{\xi}(t) = \{x(t), x(t + \tau)\} \tag{5.41}$$

$$\vec{\xi}(t + \tau) = \{x(t + \tau), x(t + 2\tau)\} .$$

Since the components of $\vec{\xi}$ are related to $\{x(t), y(t)\}$ via the one-to-one relationships

$$\xi_1(t) = x(t) \tag{5.42a}$$

$$\xi_2(t) = x(t + \tau) = \int_t^{t+\tau} dt' F_1 \{x(t'), y(t')\} + x(t) \cong$$

$$\cong \tau F_1 \{x(t), y(t)\} + x(t) \tag{5.42b}$$

with a Jacobian $|\tau(\partial F_1/\partial y)| \neq 0$, it is plausible that the information contained in the time sequences $\vec{x}(t_i)$ and $\vec{\xi}(t_i)$ $(t_i = i\tau)$ is the same, and both sequences should lead to the same characteristic dimensions. A simple example for which $\vec{x}(t_i)$ and $\vec{\xi}(t_i)$ are indeed completely equivalent is a circle:

$$\vec{x}(t_i) = \{x(t_i), y(t_i)\} = \{\sin(2\pi t_i), \cos(2\pi t_i)\} =$$

$$= \left\{\sin(2\pi t_i), \sin\left[2\pi\left(t_i + \frac{1}{4}\right)\right]\right\} = \left\{x(t_i), x\left(t_i + \frac{1}{4}\right)\right\} = \vec{\xi}(t_i) . \tag{5.43}$$

But we should be aware that arguments are only heuristic and can only be applied "cum grano salis" to situations where strange attractors appear. What Takens (1981) actually proved is the following: "If $\dot{\vec{x}} = \vec{F}(\vec{x})$ generates a d-dimensional flow, then eq. (5.44),

$$\vec{\xi}(t) = \{x_j(t), \quad x_j(t + \tau), \quad \ldots x_j[t + (2d + 1)\tau]\} \tag{5.44}$$

where $x_j(t)$ is an arbitrary component of $\vec{x}$, provides a smooth embedding for this flow, and the metric properties in both spaces (the d-dimensional $\{\vec{x}(t)\}$ and the $(2d + 1)$-dimensional $\{\vec{\xi}(t)\}$) are the same in the sense that distances in $\{\vec{x}(t)\}$ and $\{\vec{\xi}(t)\}$ have a ration which is uniformly bounded and bounded away from zero".

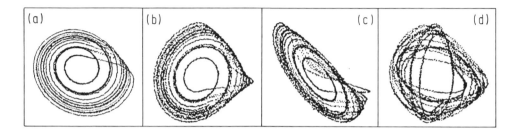

Fig. 78: Reconstruction of the Rössler attractor for $a = 0.15$, $b = 0.20$, $c = 10.0$; $\vec{x}_0 = (10.0, 0, 0)$ from a time series: a) x, y coordinates of the "true" attractor obtained by numerical integration of eq. (5.45 a–c); b) and c) reconstructions for $\tau = 0.23, 0.39, 3.26$, measured in units of the average orbital time, respectively. (After Fraser and Swinney, 1986.)

Fig. 78 shows a reconstruction of (a projection of) the Rössler attractor (Rössler, 1976), which is generated by the system,

$$\dot{x} = -z - y \tag{5.45a}$$

$$\dot{y} = x + ay \tag{5.45b}$$

$$\dot{z} = b + z(x - c) \tag{5.45c}$$

from about $6 \cdot 10^5$ points for different choices of the delay time τ. Although for an infinite amount of noise free data, τ could be chosen almost arbitrary (Takens, 1981), it can be seen from Fig. 78 that for a finite time series the quality of the reconstruction depends on τ. If τ is too small, $x(t)$ and $x(t + \tau)$ become practically indistinguishable and one obtains a linear dependence that is not present for the coordinates of the real trajectory. It is, therefore, reasonable to choose the decay time of the autocorrelation function $C(t)$ of the signal x_n for τ

$$C(t) = \lim_{N \to \infty} \frac{1}{N} \sum_{n=1}^{N-1} x_n x_{n+1} \equiv \langle x_0 x_t \rangle \tag{5.46a}$$

$$C(\tau) \approx \frac{1}{2} C(0) \tag{5.46b}$$

which ensures that $x(t)$ and $x(t + \tau)$ become linearly independent, but other choices for τ have also been proposed (Fraser and Swinney, 1986; Liebert, Kaspar and Schuster, 1987).

Generalized Dimensions and Distribution of Singularities in the Invariant Density

In this section, we discuss the meaning of the generalized dimensions D_q for special values of q and demonstrate explicitly the connection of D_q to the distribution $f(\alpha)$ of singularities in the invariant density of a strange attractor. Proceeding in a similar way as in Section 5.2, we chop the trajectory $\vec{x}(t) = [x_1(t) \ldots x_d(t)]$ of a dynamical system on a strange attractor into a sequence of points $\vec{x}(t = 0)$, $\vec{x}(t = \tau) \ldots \vec{x}(t = N\tau)$ and partition the d-dimensional phase space into cells l^d. The probability p_i of finding a point of the attractor in cell number i ($i = 1, 2 \ldots M(l)$) is then given by

$$p_i = \lim_{N \to \infty} \frac{N_i}{N} \tag{5.47}$$

where N_i is the number of points $\{\vec{x}(t = j\tau)\}$ in this cell.

The generalized dimensions D_q which are related to the qth powers of p_i via

$$D_q = \lim_{l \to 0} \frac{1}{q-1} \frac{\log \left(\sum_{i=0}^{M(l)} p_i^q \right)}{\log l} \; ; \quad q = 0, 1, 2 \ldots \tag{5.48}$$

For $q \to 0$ we obtain from (5.48)

$$D_0 = \lim_{l \to 0} (\log \sum_{i=0}^{M(l)} 1)/\log l = -\lim_{l \to 0} \frac{\log M(l)}{\log l} \tag{5.49}$$

which is just the usual definition (3.69) of the Hausdorff dimension of the attractor (i.e. $D = D_0$).

As $q \to 1$, eq. (5.48) becomes

$$D_1 = -\lim_{l \to 0} \frac{S(l)}{\log l} \tag{5.50}$$

where

$$S(l) = - \sum_{i=0}^{M(l)} p_i \log p_i \, . \tag{5.51}$$

Since $S(l)$ is the information gained, if we know $\{p_i\}$ and learn that the trajectory is in a specific cell i, D_1 is called the information dimension. It tells us how this information gain increases as $l \to 0$.

For a homogeneous attractor where all p_i are the same, i.e. $p_i = 1/M(l)$, we have

$$S(l) = - \sum_{i=0}^{M(l)} \frac{1}{M(l)} \log \frac{1}{M(l)} = \log M(l) \, . \tag{5.52}$$

Furthermore, the information dimension is always less or equal to the Hausdorff dimension, that is,

$$D_1 \leqslant D_0 \, . \tag{5.53}$$

This can be proven by maximizing $S(l)$ under the constraint $\sum_i p_i = 1$:

$$\frac{\partial}{\partial p_j} \left[- \sum_{i=1}^{M(l)} p_i \log p_i + \lambda \sum_i p_i \right] = 0 \tag{5.54a}$$

$$\to p_j = e^{-1+\lambda} \, . \tag{5.54b}$$

After eliminating the Lagrange multiplier λ via the constraint, eq. (5.54b) yields

$$p_j = \frac{1}{M(l)} \tag{5.55}$$

$$S(l) \leqslant \max [S(l)] = \log M(l) \tag{5.56}$$

from which eq. (5.53) follows after division by $\log l$.

The inequality (5.53) has been generalized to (Hentschel and Procaccia, 1983):

$$D_{q'} \leqslant D_q \text{ for } q' > q \tag{5.57}$$

where the equality sign holds if the attractor is uniform.

In order to explain the connection between the D_q's and the singularities in the invariant density of an attractor, we calculate D_q for a one dimensional system which has a power law singularity in its invariant density $\rho(x)$ at $x = 0$, i.e.

$$\rho(x) = \frac{1}{2} x^{-\frac{1}{2}} \quad \text{for} \quad x \in [0, 1] \, . \tag{5.58}$$

This is just the behavior of $\rho(x)$ near $x = 0$ for the logistic map at $r = 4$ (see eq. (3.106) where we ignored the singularity at $x = 1$ to simplify our argument. (The D_q's will be the same for both systems).

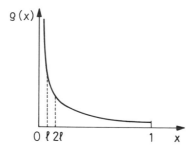

Fig. 79: The invariant density $\rho(x) = 1/2 (x)^{-\frac{1}{2}}$ for $0 \leqslant x \leqslant 1$, $x_1 = 0$, $x_2 = l$, ... (schematically).

Fig. 79 shows that:

$$p_i = \int\limits_{x_i}^{x_i+l} \rho(x)\,dx \propto \begin{cases} l^{\frac{1}{2}} & \text{for} \quad i = 1 \\ l^1 & \text{for} \quad i \neq 1 \end{cases} \tag{5.59}$$

Thus,

$$\sum_i p_i^q = \left[\int_0^l \rho(x)\,dx\right]^q + \sum_{i \neq 1} [\rho(x_i)\,l]^q \cong \tag{5.60a}$$

$$\cong \left[\int_0^l \rho(x)\,dx\right]^q + l^{q-1}\int_l^1 dx\,\rho(x)^q = \tag{5.60b}$$

$$= (1 - a)\,l^{\frac{q}{2}} + al^{q-1} \tag{5.60c}$$

where $a = \left(\dfrac{1}{2}\right)^q \left(1 - \dfrac{q}{2}\right)^{-1}$. Eq. (5.60c) can be written as

$$\sum_i p_i^q \cong \int d\alpha\,\rho(\alpha)\,l^{-f(\alpha)}\,l^{\alpha q} \tag{5.61}$$

where

$$\rho(\alpha) = (1 - \alpha)\,\delta\left(\alpha - \frac{1}{2}\right) + a\delta(\alpha - 1) \tag{5.62}$$

and

$$f(\alpha) = \begin{cases} 0 & \text{for} \quad \alpha = \dfrac{1}{2} \\ 1 & \text{for} \quad \alpha = 1 \end{cases}. \tag{5.63}$$

The interpretation of eqns. (5.59–63) is as follows. Associated with different singularities for the density of states $\rho(x)$ (e.g. $\rho(x \to 0) \sim x^{-\frac{1}{2}}$, $\rho(x) \sim$ const. otherwise) which gives rise to different singularities for $p_i(l) \sim l^{\alpha_i}$ in eq. (5.59), are different exponents $f(\alpha)$ which measure the fractal dimension of the density of these singularities on the attractor. The singularity with exponent $\alpha = 1/2$ occurs just at one point, that is, $f(\alpha = 1/2) = 0$, whereas $\alpha = 1$ occurs in a whole one dimensional interval, that is $f(\alpha = 1) = 1$. This concept of a distribution $f(\alpha)$ of fractal dimensions which are associated with a whole set of singularities of strength α can be generalized to strange attractors, and it turns out that the functions $f(\alpha)$ are again universal, for example, for the Feigenbaum attractor or the attractor, which is associated with the transition from quasiperiodicity to chaos. This type of universality, which is again associated with a whole set of singularities and not just a single exponent, is also called global universality. (It should be noted that the term global universality is also used if a whole range of parameter values has universal properties as will be explained in Chapter 6).

By way of generalizing our example, we now make the scaling hypothesis that

$$p_i(l) \sim l^{\alpha_i} \tag{5.64}$$

(where i denotes the box i of linear dimension l) occurs in $\sum_i p_i^q$ with a density

$$\rho(\alpha) l^{-f(\alpha)} \, d\alpha \tag{5.65}$$

such that the sum can be estimated as:

$$Z_q \equiv \sum_i p_i^q \cong \int d\alpha \, \rho(\alpha) l^{-f(\alpha)} l^{\alpha q} = \int d\alpha \, \rho(\alpha) e^{[-f(\alpha)+\alpha q] \log l} \tag{5.66}$$

i.e. we wrote Z_q as an integral over the singularities α. In the limit $l \to 0$, the integral can be evaluated using the saddle point approximation (see Appendix E, eq. (E. 7)) and becomes:

$$Z_q \propto e^{-[f(\alpha)-\alpha q] \log l} \tag{5.67}$$

where the dominating value of α is determined by:

$$\frac{\partial}{\partial \alpha} [q\alpha - f(\alpha)] = 0 \to f'(\alpha) = q \tag{5.68a}$$

$$\frac{\partial^2}{\partial \alpha^2} [q\alpha - f(\alpha)] > 0 \to f''(\alpha) < 0. \tag{5.68b}$$

This yields via eqns. (5.33) and (5.66–67) for D_q:

$$D_q = \{q\alpha(q) - f[\alpha(q)]\}/(q-1) \tag{5.69}$$

and after differentiation

$$\alpha(q) = \frac{\partial}{\partial q}[(q-1)D_q].$$ (5.70)

By eliminating, via eq. (5.70), the variable q in favor of α and using eq. (5.69), one obtains $f(\alpha)$ as (negative) Legendre transformation of $(q-1)D_q$:

$$f(\alpha) = q(\alpha) - [q(\alpha) - 1]D_{q(\alpha)}.$$ (5.71)

For our example with $\rho(x) = x^{-\frac{1}{2}}$, we find from eqns. (5.69–71):

	D_q	$\alpha(q)$	$f(\alpha)$
$q \geqslant 2$	$\dfrac{1}{2}\dfrac{q}{q-1}$	$\dfrac{1}{2}$	0
$q \leqslant 2$	1	1	1

i.e. the α-spectrum consists of two points, as calculated above (see eq. (5.63)).

Numerical determination of the dimensions D_q, by covering the phase space with a set of boxes of volume l^d and counting the number of iterates which lie in a certain cell, is rather cumbersome and in fact impossible for attractors of higher dimensions. However, we can replace the sum over the uniformly distributed boxes in $\sum\limits_i p_i^q$ by a sum over nonuniformly distributed boxes around the points x_j of a time series which results e.g. from a map $x_{j+1} = f(x_j)$.

$$\sum_i p_i^q = \sum_i \left[\int_{\text{Box } i} \rho(x)\,dx\right]^q \cong$$

$$\cong \sum_i [\rho(x_i)\,l]^q = \sum_i \rho(x_i)\,l\,[\rho(x_i)\,l]^{q-1} \cong$$

$$\cong \int\rho(x)\,dx\,\tilde{p}(x)^{q-1} \cong \frac{1}{N}\sum_j \{\tilde{p}[f^j(x_0)]\}^{q-1} =$$

$$= \frac{1}{N}\sum_j \tilde{p}_j^{q-1}.$$ (5.72)

Here x_i is an element of box i and $\tilde{p}[f^j(x_0)] \equiv \tilde{p}_j$ is the probability of the trajectory to be in a box of size l around the iterate $x_j = f^j(x_0)$. Eq. (5.72) should make it plausible, mathematical rigor is not attempted, that the change from p_i^q to $\tilde{p}_j^{q-1}$ is due to the fact that the points x_j of the time series (and the boxes around them) are nonuniformly distributed. We next generalize eq. (5.72) to higher dimensional systems and write the probability $\tilde{p}_j$ that an element of the time series falls into an interval l around the element $\vec{x}_j$ as:

$$\tilde{p}_j = \frac{1}{N} \sum_i \Theta \left(l - |\vec{x}_i - \vec{x}_j| \right) \tag{5.73}$$

where $\Theta(x)$ is the Heaviside step function.

Using eqns. (5.72–73), $\sum_i p_i^q$ becomes

$$\sum_i p_i^q = \frac{1}{N} \sum_j \left[\frac{1}{N} \sum_i \Theta(l - |\vec{x}_i - \vec{x}_j|) \right]^{q-1} = C^q(l) . \tag{5.74}$$

For $q = 2$, this reduces to the correlation integral $C(l)$ introduced by Grassberger and Procaccia (1983 a) which measures the probability of finding two points of an attractor in a cell of size l:

$$\sum_{i=0}^{M(l)} p_i^2 = \text{the probability that two points of the attractor lie within a cell } l^d$$

$$\simeq \text{the probability that two points at the attractor are separated by a distance smaller than } l$$

$$= \lim_{N \to \infty} \frac{1}{N^2} \{\text{number of pairs } ij \text{ whose distance } |\vec{x}_i - \vec{x}_j| \text{ is less than } l\}$$

$$= \lim_{N \to \infty} \frac{1}{N^2} \sum_{ij} \theta(l - |\vec{x}_i - \vec{x}_j|)$$

$$= C(l) = \text{correlation integral} . \tag{5.75}$$

The correlation integral $C(l)$ can be used to determine the following properties from a measured time series:

— *The correlation dimension D_2:*

$$D_2 = \lim_{l \to 0} \frac{1}{\log l} \log \sum_i p_i^2 \tag{5.76}$$

which yields a lower bound to the Hausdorff dimension D_0 i.e. $D_2 < D_0$. Fig. 80 shows how D_2 is determined from $C(l)$ for the Hénon map.

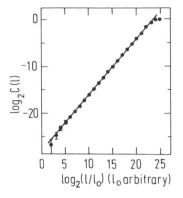

Fig. 80: $\log_2 C(l)$ versus $\log_2 l$ for the Hénon map. The slope yields $D_2 = 1.21$ (after Grassberger and Procaccia, 1983a).

Fig. 81 demonstrates how Takens' reconstruction of the trajectory from the measurement of a single variable (Takens, 1983) works for the computation of $C(l)$ for the Lorenz attractor.

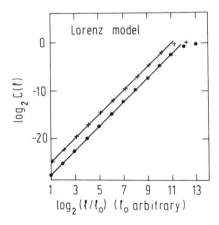

Fig. 81: $\log C(l)$ versus $\log l$ for the Lorenz model. The lower line is obtained directly from the three-dimensional time series $\{x(t_i), y(t_i), z(t_i)\}$, whereas the upper line originates from the reconstructed series $\vec{\xi}(t_i) = \{x(t_i), x(t_i + \tau), x(t_i + 2\tau)\}$. The slopes of both curves are the same, i.e. the correlation dimension $D_2 = 2.05$ obtained by both methods is the same, as stated above (Grassberger and Procaccia, 1983a).

— *The embedding dimension d:*

Fig. 82 shows the l dependence of the correlation integral for the Mackey-Glass system. Although this system has an infinite dimension, its correlation dimension is finite and smaller than 3. It is therefore sufficient to use a simple time series with a three-dimensional vector $\vec{\xi}(t_i) = \{x(t_i), x(t_i + \tau), x(t_i + 2\tau)\}$ to determine D_2. The dimension d in $\vec{\xi}(t) = \{x(t_i) \ldots x(t_i + (d-1)\tau)\}$, above which D_2 no longer changes, is the (minimal) *embedding dimension* of the attractor.

— *Separation of deterministic chaos and external white noise:*

The correlation integral can also be used as a tool to *distinguish between deterministic irregularities,* which arise from intrinsic properties of the strange attractor, *and external white noise.* Suppose we have a strange attractor embedded in

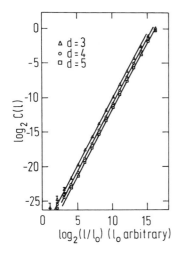

Fig. 82: $D_2 = 1.95 \pm 0.03$ determined from a single-variable time series for the Mackey-Glass equation with parameter values $\tau = 17$, $a = 0.2$, $b = 0.1$ for different embedding dimensions d (Hentschel and Procaccia, 1983.)

d-dimensional space and we add an external white noise. Each point on the attractor then becomes surrounded by uniform d-dimensional cloud of points. The radius of this cloud is given by the noise amplitude l_0. For $l \gg l_0$, eq. (5.74) counts these clouds as points, and the slope of a plot of log $C(l)$ versus log l yields the correlation exponent of the attractor. For $l \ll l_0$ most of the points counted lie within the uniformly filled d-dimensional cells, and the slope crosses over to d, as shown in Fig. 83 for the noisy Hénon attractor.

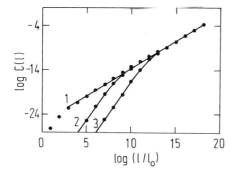

Fig. 83: log $C(l)$ as a function of log l for the Hénon map embedded in three dimensions. Curve 1 is for the map without noise and yields $D_2 = 1.25$. Curve 2 is for the map with random noise with amplitude $5 \cdot 10^{-3}$. Curve 3 is for the map with noise amplitude $5 \cdot 10^{-2}$. Curve 2 and 3 break at length scales that are determined by the noise level below which the slope is approximately 3 (Ben-Mizrachi et al., 1983.)

Finally, let us briefly comment on the intuitive meaning of the variable q and then present two examples of D_q and $f(\alpha)$ curves.

If one replaces, in the definition of D_q via eq. (5.33), the p_i by the probabilities $\tilde{p}_j$, for a trajectory to fall in a box around an iterate (see eq. 5.72) then the resulting expression

$$D_q = -\lim_{l \to 0} \frac{1}{q-1} \left| \frac{1}{\log l} \right| \sum_j \tilde{p}_j(l)^{q-1} \qquad (5.77)$$

resembles closely the expression F_β of the free energy of an N-particle equilibrium system at a temperature $T = \beta^{-1}$:

$$F_\beta = -\lim_{N \to \infty} \frac{1}{\beta} \cdot \frac{1}{N} \sum_i (e^{-E_i})^\beta . \tag{5.78}$$

The variable $|\log l|$ corresponds to the number of particles and $q - 1$ corresponds to the inverse temperature β. It follows already from eq. (5.77) that, for, $q \to +\infty$, the most concentrated parts of the measure (large $\tilde{p}_i$'s) are being stressed; whereas for $q \to -\infty$, the most rarified parts (small $\tilde{p}_i$'s become dominant. In this sense, q indeed serves as the (inverse) temperature in statistical mechanics where at every temperature a different set of energy levels E_i (i. e. probabilities $\exp(-\beta E_i)$) becomes dominant in the free energy.

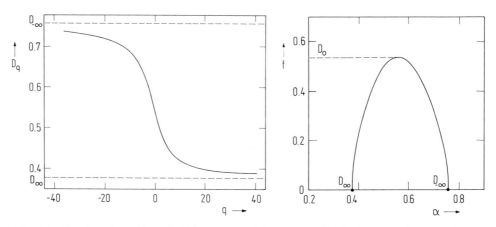

Fig. 84: The functions D_q and $f(\alpha)$, computed from eqns. (5.33), (5.71), and (5.74), for the Feigenbaum attractor (after K. Pawelzik, priv. comm.).

Fig. 84 shows the D_q and $f(\alpha)$ curves for the Feigenbaum attractor that is generated by the iterates of the logistic map $x_{n+1} = r x_n (1 - x_n)$ at $r = r_\infty = 3.5699$ ... (see sect. 3.4). The function $f(\alpha)$ must be concave because eq. (5.68b) requires $f''(\alpha) < 0$, and the maximum of $f(\alpha)$ at $\alpha = \alpha_m$ is equal to the Hausdorff dimension D_0 because at the maximum $f'(\alpha_m) = 0$ which yields via eqns. (5.68a–b)

$$f'(\alpha_m) = q_m = 0 \tag{5.79}$$

and

$$D_0 = f(\alpha_m) . \tag{5.80}$$

Furthermore, we see from eq. (5.69) that, as long as $f(\alpha)$ remains bounded, the limiting dimensions $D_{+\infty}$ becomes equal to the corresponding α values, i.e. $D_{+\infty} = \alpha(+\infty)$ which implies via eq. (5.59) $f[\alpha(+\infty)] = 0$. Thus, the zeros of $f(\alpha)$ are equal to $D_{\pm\infty}$, and the slope of $f(\alpha)$ is infinite at these points because of eq. (5.68a).

The dimension $D_{-\infty}$ which is associated with the most rarified regions of the Feigenbaum attractor can be calculated as follows. The size l_n of the most rarified region on the 2^n attractor, which approaches the Feigenbaum attractor for $n \to \infty$, decreases as α^{-n} where α is the Feigenbaum constant. This is due to the fact that the function $\sigma(x)$ from sect. 3.3 which measures the ratio of the distances between the elements of subsequent supercycles, has its maximum at α^{-1} (see Fig. 31); i.e., the largest distance decreases like α^{-n}. The probability p_n of a point on the 2^n cycle to lie within the interval l_n is just $p_n = 2^{-n}$ because only one point of the cycle is contained in l_n. Putting everything together, $D_{-\infty}$ becomes:

$$D_{-\infty} = \lim_{q \to -\infty} \lim_{n \to \infty} \frac{1}{q-1} \frac{1}{\log l_n} \log p_n^{q-1} = \frac{\log 2}{\log \alpha} \cong 0.75551 \ldots \quad (5.81)$$

which is in excellent agreement with the numerical result in Fig. 84 obtained from the time series of the logistic map. Fig. 84 shows that D_q converges very slowly against its limits $D_{+\infty}$, but $\alpha(q = +\infty) = D_{+\infty}$ can be easily extrapolated from the corresponding $f(\alpha)$ curves. Thus, the transformation to $f(\alpha)$ leads to better estimates of $D_{+\infty}$ than the direct calculation of the D_q's. Another advantage of the $f(\alpha)$ spectrum is the fact that it represents (e. g. for the Feigenbaum attractor) a smooth universal curve which yields the global density of scaling indices. The universal function $\sigma(x)$ of Feigenbaum, which everywhere describes the local scaling (see section 3.3), contains in principle the same (and even more) information as $f(\alpha)$, but it is nowhere differentiable and, is, therefore, a function that is difficult to use. A further example where the merits of the $f(\alpha)$ representation of experimental data become obvious is given in chapter 6 where we investigate the question whether an experimental orbit obtained from a forced Rayleigh-Bénard experiment is in the same universality class as the orbit generated from a circle map.

Generalized Entropies and Fluctuations around the K-Entropy

We generalize in this section the expression

$$K = \lim_{l \to 0} \lim_{n \to \infty} \frac{1}{n} \sum_{i_0 \ldots i_{n-1}} P_{i_0 \ldots i_{n-1}} \log P_{i_0 \ldots i_{n-1}} \quad (5.82)$$

for the Kolmogorov entropy of a map (see eq. 5.15) by introducing in analogy to the D_q's a whole set of entropies K_q:

$$K_q = - \lim_{l \to 0} \lim_{n \to \infty} \frac{1}{n} \frac{1}{q-1} \log \sum_{i_0 \ldots i_{n-1}} P^q_{i_0 \ldots i_{n-1}} \qquad (5.83)$$

and we show, by way of an example, that their Legendre transformation is related to the spectrum of fluctuations $g(\lambda)$ around the K-entropy.

If we introduce a variable $T = e^{-n}$, eq. (5.83) can be rewritten as

$$K_q = \lim_{l \to 0} \lim_{T \to 0} \frac{1}{\log T} \frac{1}{q-1} \log \sum_{i_0 \ldots i_{n-1}} P^q_{i_0 \ldots i_{n-1}} \qquad (5.84)$$

which looks — apart from the fact that we have a whole series of indices, instead of just one — similar to eq. (5.23) for the D_q's with l replaced by T. It is, therefore, reasonable to try, in analogy to eq. (5.64) the scaling ansatz

$$P_{i_0 \ldots i_n} \propto T^{\lambda \, (i_0 \ldots i_n)} \qquad (5.85)$$

where the number of $\lambda \, (i_0 \ldots i_n)$ in the interval $\lambda, \lambda + d\lambda$ is (in analogy to eq. (5.65)) proportional to

$$\rho(\lambda) \, T^{-g(\lambda)} \, d\lambda \ . \qquad (5.86)$$

Using the same arguments as in eq. (5.86), we arrive at the limit $T \to 0$ at

$$K_q = \frac{1}{q-1} \, [\lambda q - g(\lambda)] \qquad (5.87)$$

where λ is determined by the saddle point conditions:

$$g'(\lambda) = q \qquad (5.88\,a)$$

and

$$g''(\lambda) < 0 \ . \qquad (5.88\,b)$$

In order to see the physical meaning of the numbers λ and of the distribution $g(\lambda)$, we consider as a simple example the piecewise expanding map

$$f(x) = \begin{cases} \dfrac{x}{p} & \text{for} \quad 0 \le x \le p \\[2mm] \dfrac{1-x}{1-p} & \text{for} \quad p \le x \le 1 \end{cases} \qquad (5.89)$$

shown in Fig. 85 and compute K_q and $g(\lambda)$ explicitly for the dynamical system defined by

$$x_{n+1} = f(x_n) . \qquad (5.90)$$

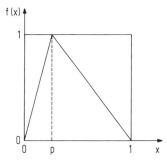

f(x)

Fig. 85: Tent map defined via eq. (5.89).

The probabilities $P_{i_0 \ldots i_{n-1}}$ and the sums $\sum\limits_{i_0 \ldots i_{n-1}} P^q_{i_0 \ldots i_{n-1}} \equiv S^n_q$ appearing in eq. (5.84) then become:

$$P_{i_0} = \begin{cases} p & \text{if } x_1 \in [0, p] \\[2mm] 1 - p & \text{if } x_1 \in [p, 1] \end{cases} \qquad \to S^1_q = p^q + (1 - p)^q \qquad (5.91)$$

$$P_{i_0 i_1} = \begin{cases} p^2 & \text{if } x_1, x_2 \in [0,p] \\[1mm] p(1 - p) & \text{etc.} \\[1mm] (1 - p)^2 \end{cases} \qquad \to S^2_q = [p^q + (1 - p)^q]^2$$

i.e. $S^n_q = [p^q + (1 - p)^q]^n$

which yields for K_q:

$$K_q = \frac{-1}{q - 1} \log [p^q + (1 + p)^q] . \qquad (5.92)$$

For the limit $q \to 1$, we obtain from eq. (5.92) the Kolmogorov entropy

$$K_1 = p \log \left(\frac{1}{p} \right) + (1 - p) \log \left(\frac{1}{1 - p} \right) . \qquad (5.93)$$

K_1 is, as expected, equal to the positive Liapunov exponent λ_m of the system, which can also be obtained directly from

$$\lambda_m = \int dx \rho(x) \log |f'(x)| = p \log \left(\frac{1}{p}\right) + (1-p) \log \left(\frac{1}{1-p}\right) \qquad (5.94)$$

where we used the fact that the invariant density $\rho(x) = 1$ for $f(x)$. (This can be checked by using eq. (5.89) in the Frobenius-Perron equation (2.30)).

Next, we compute $\lambda(q)$ and $g(\lambda)$ via eqns. (5.87, 5.92):

$$\lambda(q) = \frac{\partial}{\partial q}(q-1)K_q =$$

$$= -[p^q \log p + (1-p)^q \log(1-p)]/[p^q + (1-p)^q] \qquad (5.95)$$

which becomes for

$$x = p^q/[p^q + (1-p)^q] \qquad (5.96)$$

equal to

$$\lambda = \lambda(x) = -[x \log p + (1-x) \log(1-p)] . \qquad (5.97)$$

Similarly, we obtain from (5.87) and (5.96):

$$g[\lambda(q)] = (q-1)K_q - q\lambda(q) =$$

$$= -\log[p^q + (1-p)^q] +$$

$$+ [p^q \log p^q + (1-p)^q \log(1-p)^q]/[p^q + (1-p)^q]$$

$$= x \log x + (1-x) \log(1-x) . \qquad (5.98)$$

Eq. (5.98) yields $g(\lambda)$ as shown in Fig. 88 if we eliminate x in favor of λ via eq. (5.97). The quantity $\lambda(x)$, which appears in eq. (5.97), has a very simple interpretation. It is just the "Liapunov exponent" of a *finite* series of iterates of length n which visits r times the interval $[0, p]$ in Fig. 85:

$$e^{n\lambda(y)} = \prod_j |f'(x_j)| = \left(\frac{1}{p}\right)^r \left(\frac{1}{1-p}\right)^{n-r} \qquad (5.99)$$

$$\rightarrow \lambda(y) = -[y \log p + (1-y) \log(1-p)] \qquad (5.100)$$

where $y = r/n$. The probability $\hat{P}$ of finding $\lambda(y)$ is equal to the probability of finding in a series of iterates, r times a point which is in $[0, p]$ and $(n-r)$ times a point which is located in $[p, 1]$, that is,

$$\hat{P} = \binom{n}{r} p^r (1 - p)^{n-r}. \tag{5.101}$$

Using Sterlings's formula $n! \simeq n^n$, this becomes:

$$\log \hat{P} \simeq n \left[y \log \left(\frac{y}{p} \right) + (1 - y) \log \left(\frac{1 - y}{1 - p} \right) \right] = \tag{5.102}$$

$$= n \left[y \log y + (1 - y) \log (1 - y) - \lambda (y) \right]$$

where we can again replace y by λ via eq. (5.90) to obtain $\hat{P}(\lambda)$.

By comparing eqns. (5.97 − 98) and (5.100, 5.102), we see that $e^{ng(\lambda)}$ is (apart from a factor e^λ) equal to the probalility $\hat{P}(\lambda)$ of seeing in a finite series of iterates the "Liapunov exponent" λ. The Legendre transformation from the variable q in K_q to the variable λ yields, therefore, the distribution $g(\lambda)$ which describes the fluctuations of the Liapunov exponent for a time series of length n. Note that we used for our interpretation a map which is piecewise expanding (i.e. $|f'(x)| \geqslant 1$ for all $x \in [0, 1]$) and which yields, therefore, only positive expansion rates λ. For general systems (which can also be higher dimensional), our results generalize to the statement that $e^{ng(\lambda)}$ describes the fluctuation spectrum of the (sum of the) positive Liapunov exponents, i.e. of the Kolmogorov entropy for finite time series (see Fig. 86).

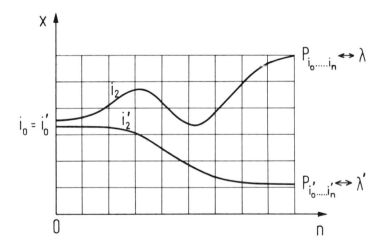

Fig. 86: Different trajectories of length n occur with probabilities $P_{i_0 \ldots i_{n-1}}$. The distribution of the finite time "K-entropies" $\lambda = -\lim\limits_{l \to 0} (1/n) \sum\limits_{i_0 \ldots i_{n-1}} P_{i_0 \ldots i_{n-1}} \log P_{i_0 \ldots i_{n-1}}$ is given by $e^{ng(\lambda)}$ where $g(\lambda)$ is the negative Legendre transformation of K_q.

The numerical computation of K_q from a measured time series proceeds in a fashion which is closely analogous to the D_q's. By generalizing eq. (5.74):

$$\sum_i p_i^q = \frac{1}{N} \sum_i \left\{ \frac{1}{n} \sum_j \Theta \left[l - |\vec{x}_i - \vec{x}_j| \right] \right\}^{q-1} = C_q(l) \qquad (5.103)$$

to a whole trajectory of length n we obtain:

$$\sum_{i_0 \ldots i_{n-1}} P_{i_0 \ldots i_{n-1}}^q = \frac{1}{N} \sum_i \left\{ \frac{1}{N} \sum_j \Theta \left[l - \sqrt{\sum_{m=0}^{n} (\vec{x}_{i+m} - \vec{x}_{j+m})^2} \right] \right\}^{q-1} \equiv \qquad (5.104)$$

$$\equiv C_n^q(l) \,.$$

This is again a generalization of a correlation integral $C_n(l)$ which has been introduced by Grassberger and Procaccia (1983 b):

$$C_n(l) = \lim_{N \to \infty} \frac{1}{N^2} \left\{ \text{number of pairs } ij \text{ with } \sqrt{\sum_{m=0}^{n-1} (\vec{x}_{i+m} - \vec{x}_{j+m})^2} < l \right\} \simeq$$

$$\simeq \sum_{i_0 \ldots i_{n-1}} P_{i_0 \ldots i_n}^2 \qquad (5.105)$$

and which yields the correlation entropy:

$$K_2 = - \lim_{l \to 0} \lim_{n \to \infty} \frac{1}{n} \log \sum_{i_0 \ldots i_{n-1}} P_{i_0 \ldots i_{n-1}}^2 \,. \qquad (5.106)$$

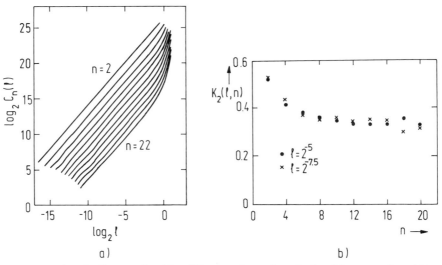

Fig. 87: a) $\log C_n(l)$ versus $\log l$ for different values of n calculated from a series of $N = 15\,000$ points for the Hénon map. b) As $n \to \infty$ and $l \to 0$, $K_2(n, l)$ calculated from eq. (5.47) approaches the common value $K_2 = 0.325 \pm 0.2$ (Grassberger and Procaccia, 1983 b).

K_2 represents, in analogy to eq. (5.57), a lower bound to the K-entropy $K = K_1 \geq K_2$, and $K_2 > 0$ provides a *sufficient* condition for chaos. Fig. 69 shows the results for $C_n(l)$ and K_2 for the Hénon map with $a = 1.4$, $b = 0.3$.

However, eqns. (5.83, 5.104) yield, for $q \to 1$, an explicit expression for the K-entropy itself:

$$K = K_1 = \lim_{l \to 0} \lim_{n \to \infty} \frac{1}{n} \frac{1}{N} \sum_i \log \left\{ \frac{1}{N} \sum_j \Theta \left[l - \sqrt{\sum_{m=0}^{n-1} (\vec{x}_{i+m} - \vec{x}_{j+m})^2} \right] \right\}$$

(5.107)

which can be calculated from a measured signal. The condition $K > 0$ provides, of course, a sharper condition for chaos than $K_2 > 0$ (see also Cohen and Procaccia, 1985).

Let us finally summarize our results by a single formula which demonstrates that all generalized dimensions D_q and entropies K_q can be extracted from experimental data. Eqns. (5.83) and (5.104) yield for $n \to \infty$ and $l \to 0$:

$$\log C_n^q(l) \propto n(q - 1) K_q .$$

(5.108)

If we watch the sequence of limits (first $n \to \infty$ then $l \to 0$), we can combine eqns. (5.33, 5.54, 5.74, 5.83, 5.104) and obtain the compact expression

$$\lim_{l \to 0} \lim_{n \to \infty} \log C_n^q(l) = (q - 1) D_q \log l + n(q - 1) K_q .$$

(5.109)

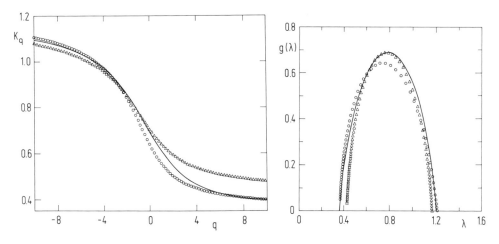

Fig. 88: a) K_q- and b) $g(\lambda)$-spectrum of the tent map (eq. (5.89)). Full lines: Theoretical curves obtained from eqns. (5.87, 5.92). Dots: Numerical results obtained via eqns. (5.87, 5.109) for 2000 iterates. (After Pawelzik and Schuster, 1987.)

Therefore, a plot of $\log C_n^q(l)$ — which can be determined from an observed time series via eq. (5.104) — versus $\log l$ yields, for fixed q and different values of n, straight lines, with slopes $(q - 1) D_q$, whose separations along the y-axis converge for $n \to \infty$ to $(q - 1) K_q$ (see, e.g., Fig. 87 for $q = 2$). The spectra $f(a)$ and $g(\lambda)$ can be obtained by Legendre transformation from these quantities. Fig. 88 shows examples of K_q and $g(\lambda)$ curves that have been obtained by this method from a numerically generated time series of the tent map (5.89).

Let us finally add a word of caution. It is by no means completely straightforward to obtain, from an experimentally measured time series, the D_q and K_q curves, because the signal is noisy, the length of the series is finite, and the delay time which is needed to reconstruct the attractor (see eq. 5.44) is generally unknown. All this adds a good deal of ambiguity to the application of the procedures described by eq. (5.109). We would like to call attention to the Proceedings of a conference on "Dimensions and Entropies in Chaotic Systems" (edited by Mayer-Kress, 1986) where merits and limits of different numerical procedures to extract dimensions, entropies, and Liapunov exponents from a time series are discussed.

Kaplan-Yorke Conjecture

Although we above made a distinction between dynamic properties of a strange attractor, such as the Liapunov exponents, and static properties measured by the D_q's, both quantities are in fact connected. For example, if we have a flow in three-dimensional phase space with two negative Liapunov exponents, we know that the attractor contracts to a line with $D_q = 1$ for all q (see Fig. 89).

Another example is the attractor which belongs to the nonarea preserving baker's transformation (5.7a, b). Its Hausdorff dimension D_B (see eq. (5.10)) can be expressed in terms of the Liapunov exponents $\lambda_1 = \log 2$, $\lambda_2 = \log a$:

$$D_B = 1 + \frac{\lambda_1}{|\lambda_2|} . \tag{5.110}$$

Kaplan and Yorke (1979) conjectured the following more general formula for arbitrary strange attractors:

$$D_{KY} = j + \frac{\sum\limits_{i=1}^{j} \lambda_i}{|\lambda_{j+1}|} . \tag{5.111}$$

Here D_{KY} is the Hausdorff dimension according to Kaplan and Yorke, and the Liapunov exponents are ordered $\lambda_1 > \lambda_2 > \ldots > \lambda_d$, such that j is the largest integer for which $\sum\limits_{i=1}^{j} \lambda_i > 0$. Although this formula has been checked numerically and

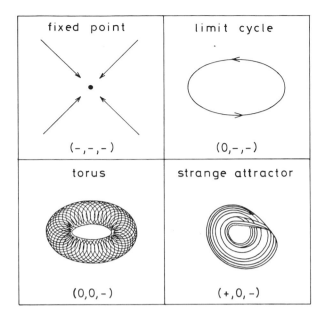

Fig. 89: Connection between the dimensions of simple attractors embedded in three-dimensional phase space and the signs of their three Liapunov exponents given in the brackets. (Zero means that the Liapunov exponent has this value.) (After Shaw, 1981.)

shown to hold for some cases by Russel et al. (1980) (see Table 9), it seems to be rigorously valid only for homogenous attractors, and its range of applicability is still an active field of research.

Table 9: Test of the Kaplan-Yorke Conjecture.

System	D (numerically)	D_{KY}
Hénon map		
$a = 1.2, b = 0.3$	1.202 ± 0.003	1.200 ± 0.001
$a = 1.4, b = 0.3$	1.261 ± 0.003	1.264 ± 0.002
Zaslavsky map		
eq. (1.12a, b)		
for $f(x) = \cos x$	1.380 ± 0.007	1.387 ± 0.001

5.4 Pictures of Strange Attractors and Fractal Boundaries

D. Ruelle writes at the end of his article on strange attractors in "The Mathematical Intelligencer" (1980): "I have not (yet) spoken of the esthetic appeal of strange attractors. These systems of curves, these clouds of points suggest sometimes fireworks or galaxies, sometimes strange and disquieting vegetal proliferations. A realm lies here to explore and harmonies to discover". Fig. 90 shows several examples of strange attractors that support this statement.

But we will see in the following that *already the boundaries of attraction* of simple rational maps of the complex plane onto iself *can have very complicated structures*. If these objects are plotted in color they show striking parallels to some of the self-similar pictures of M. C. Escher.

Let us begin with a study of the basins of attraction for the fixed points $z^* = (1, e^{2\pi i/3}, e^{4\pi i/3})$ of the map

$$z_{n+1} = z_n - (z_n^3 - 1)/(3 z_n^2) \tag{5.112}$$

in the complex plane. (Eq. (5.112) is just Newton's algorithm for the solution of $f(z) = z^3 - 1 = 0$ $(0 = f(z) \approx f(z_0) + f'(z_0)(z - z_0) \rightarrow z_1 = z_0 - f(z_0)/f'(z_0)$, etc.).)

One could think that the different basins of attraction for the roots z^* on the unit circle would be separated by straight lines, But, if one runs eq. (5.112) on a computer and colors starting points, which move to 1, $e^{2\pi i/3}$, $e^{4\pi i/3}$, in red, green and blue, respectively (and black if the starting point does not converge), one sees from the results in Plate VIII (the color Plates I–XV are shown at the beginning of the book) that the boundary of the different basins forms highly interlaced self-similar structures (see also Fig. 91). This fractal boundary solves the nontrivial problem of how to paint a plane with three colors in such a way that each boundary point of a colored region (e. g. red) is also a boundary point of the other regions (green, blue).

The boundary of a basin of attraction of a rational map is nowadays called the *Julia set* (Julia, 1918) (for a more precise definition see, e. g., Brolin, 1965). "Usually" Julia sets are fractals (for $f(z) = z^2$ the Julia set is the unit circle), and the motion of iterates on these sets is chaotic.

Next we consider the map

$$z_{n+1} = f_c(z_n) \equiv z_n^2 + c \tag{5.113}$$

in the complex plane for complex parameter values c. (Eq. 5.113) is the logistic map $x_{n+1} = rx_n(1 - x_n)$ in new variables $x = 1/2 - z/r$; $c = (2r - r^4)/4$.

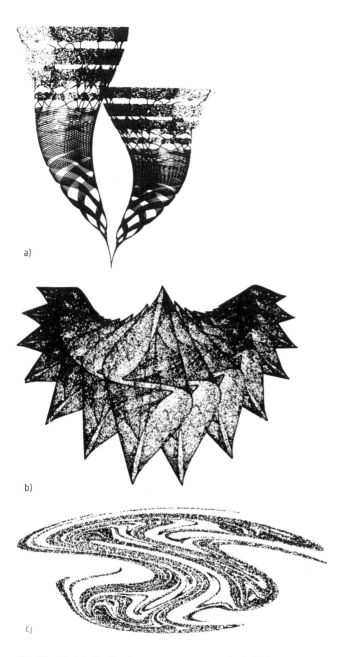

a)

b)

c)

Fig. 90: a), b), Both pictures are composed of different parts of strange attractors which arise if one iterates discretized versions of $\dot{y} = y\,(1 - y)$ and the pendulum equation, respectively (after Prüfer, 1984; Peitgen and Richter, 1984). c) Poincaré plot ($\vec{x}_n = \vec{x}\,(t = n\,T)$ of trajectories of the driven Duffing oscillator ($\ddot{x} + \gamma\dot{x} + ax + bx^3 = A + B\cos\,(2\pi t/T)$) in the chaotic régime (after Kawakami, 1984).

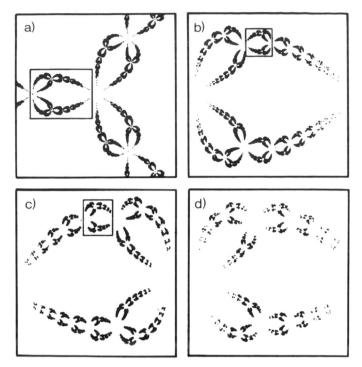

Fig. 91: Self-semilarity of the Julia set for eq. (5.112) (see also Plate VIII) (after Peitgen and Richter, 1984).

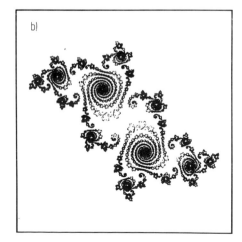

Fig. 92: Two typical Julia sets of $f_c(z)$ in eq. (5.90). a) $c = 0.32 + 0.043\,i$, b) $c = -0.194 + 0.6557\,i$. (After Peitgen and Richter, 1984.)

5   *Strange Attractors in Dissipative Dynamical Systems*

The boundary of the basin of attraction of $z^* = \infty$ forms a Julia set J_c of $f_c(z)$, which depends on c:

$$J_c = \text{boundary of } \{z \mid \lim_{n \to \infty} f_c^n(z) \to \infty\} . \tag{5.114}$$

Fig. 92 shows several examples of these sets. An important therorem by Julia (1981) and Fatou (1919) states that J_c is connected, if and only if, $\lim_{n \to \infty} f_c^n(0) \not\to \infty$. Since this limit depends only on c, one is led to consider the set M of *parameter values* c in the complex plane for which J_c is connected, i. e.

$$M = \{c \mid J_c \text{ is connected}\} = \{c \mid \lim_{n \to \infty} f_c^n(0) \not\to \infty\} . \tag{5.115}$$

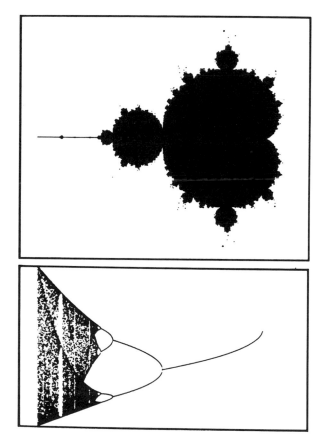

Fig. 93: Correspondence between the structure of "Mandelbrot's set" M in the c-plane and the structure of bifurcations of the (transformed) logistic map $x_{n+1} = x_n^2 + c$ along the real c-axis (after Peitgen and Richter, 1984).

The set M is called "Mandelbrot's set" after B. B. Mandelbrot who first published (1980) a picture of M (see Fig. 93). It shows that M has also a fractal structure (but it is no Julia set). This study was extended by Peitgen and Richter (1984). If c does not belong to M, then $\lim_{n \to \infty} f_c^n(0) \to \infty$. Therefore, they define "level curves" in the following way: color a starting point according to the number of iterations it needs to leave a disk with a given radius R. As shown by Douady and Hubbard (1982), lines of equal color can be interpreted as equipotential lines if the set M is considered to be a charged conductor. Plates VIII–XV show the fascinating results of this procedure which brings us back to Ruelles' remark at the beginning of this section.

6 The Transition from Quasiperiodicity to Chaos

In the first section of this chapter, we shall discuss the emergence of a strange attractor in the Ruelle-Takens-Newhouse route to turbulence (in time) and present some experimental support for this route. The subsequent section contains a study of the universal properties of the transition from quasiperiodicity to chaos via circle maps and we introduce two renormalization schemes, which are appropriate to describe local and global universality. In section 6.3, we present experimental evidence that circle maps indeed provide a useful description of the transition from quasiperiodicity to chaos in real systems. The chapter ends with a critical review of different transition scenarios that lead to a chaotic behavior.

6.1 Strange Attractors and the Onset of Turbulence

We come now to one of the most fascinating and difficult questions; namely, how the onset of fluid turbulence in time (we will not consider the distribution of spatial inhomogeneities) is related to the emergence of a strange attractor.

To understand what has been undertaken in this area, we first introduce the Hopf bifurcation (Hopf, 1942).

Hopf Bifurcation

A simple Hopf bifurcation generates a limit cycle starting from a fixed point. For example, consider the following differential equations in polar coordinates:

$$\frac{dr}{dt} = -(\Gamma r + r^3); \quad \Gamma = a - a_c \tag{6.1a}$$

$$\frac{d\theta}{dt} = \omega. \tag{6.1b}$$

Their solutions are

$$r^2(t) = \frac{\Gamma r_0^2 e^{-2\Gamma t}}{r_0^2(1 - e^{-2\Gamma t}) + \Gamma} \qquad \text{with} \quad r_0 = r(t = 0) \qquad (6.2\,\text{a})$$

$$\theta(t) = \omega t \qquad \text{with} \quad \theta(t = 0) = 0 . \qquad (6.2\,\text{b})$$

For $\Gamma \geq 0$ the trajectory approaches the origin (fixed point), whereas for $\Gamma < 0$ it spirals towards a limit cycle with radius $r_\infty = |(a - a_c)|^{1/2}$, as shown in Fig. 94.

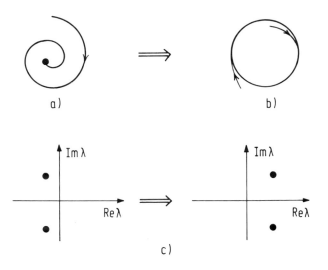

Fig. 94: Hopf bifurcation from a fixed point (a) to a limit cycle (b), and behavior of the eigenvalues λ (c).

If (6.1 a, b) is transformed into rectangular coordinates

$$\frac{dx}{dt} = -\{\Gamma + (x^2 + y^2)\}x - y\omega \qquad (6.3\,\text{a})$$

$$\frac{dy}{dt} = -\{\Gamma + (x^2 + y^2)\}y + x\omega \qquad (6.3\,\text{b})$$

and linearized about the origin, we obtain

$$\frac{d\vec{f}}{dt} = A\vec{f} \qquad (6.4)$$

where $\vec{f} = (\Delta x, \Delta y)$, and A is the matrix

$$A = \begin{pmatrix} -\Gamma & -\omega \\ \omega & -\Gamma \end{pmatrix} \tag{6.5}$$

with eigenvalues $\lambda_\pm = -\Gamma \pm i\omega$. This means that at a Hopf bifurcation a pair of conjugate eigenvalues crosses the imaginary axis, as indicated in Fig. 94c.

Landau's Route to Turbulence

A Hopf bifurcation introduces a new fundamental frequency ω into the system. As early as 1944 Landau therefore suggested a route to turbulence (in time) in which the chaotic state is approached by an infinite sequence of Hopf instabilities, as shown in Fig. 95.

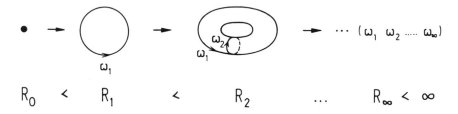

Fig. 95: Landau's route to chaos. As the parameter R increases, more and more fundamental (i.e. incommensurate) frequencies are generated by Hopf bifurcations.

Although this route leads to a time dependence which becomes more and more complicated as more and more frequencies appear, the power spectrum always remains discrete and approaches the continuum limit only after an infinite sequence of Hopf bifurcations.

Ruelle-Takens-Newhouse Route to Chaos

Fig. 96 shows that this is not the case for the Bénard experiment. After the appearance of two fundamental frequencies, the power spectrum becomes continuous.

This experiment was in fact performed, after the theoretical work of Ruelle, Takens, and Newhouse (1978) who had suggested a route to chaos which is much shorter than that proposed by Landau (1944). They showed that, after three Hopf bifurcations, regular motion becomes highly unstable in favor of motion on a strange attractor (see Fig. 97).

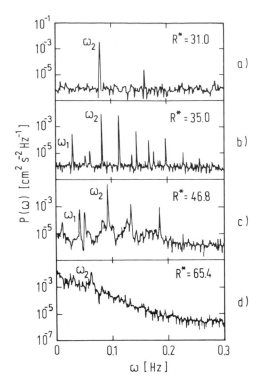

Fig. 96: Power spectrum of the convection current for a Bénard experiment (after Swinney and Gollub, 1978). With increasing (relative) Rayleigh number $R^* = R/R_c$, the following states are observed: a) periodic movement with one frequency and its harmonics, b) quasiperiodic motion with two incommensurate frequencies and their linear combinations, c) nonperiodic chaotic motion with some sharp lines, d) chaos.

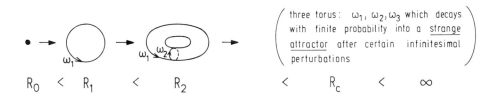

Fig. 97: The Ruelle-Takens-Newhouse route to chaos.

To be precise, we quote their theorem verbatim (Ruelle, Takens, Newhouse) (1978): „Let v be a constant vector field on the n-torus $T^n = R^n/Z^n$. If $n > 3$, every C^2 neighborhood of v contains a vector field v' with a strange Axiom A attractor. If $n > 4$, we may take C^∞ instead of C^2. (Here C^2 means that the neighborhood of the vector field is twice continuously differentiable, an Axiom A attractor (see Smale, 1967) is essentially our strange attractor, and we finally mention that the original work of Ruelle and Takens (1971) described the decay of a four torus instead of a three torus as in the theorem above).

This means practically that if a system undergoes three Hopf bifurcations, starting from a stationary solution as a parameter is varied, then it is "likely" that the system possesses a strange attractor after the third bifurcation. The power spectrum of such a system will exhibit one, then two, and then possibly three independent frequencies. When the third frequency is about to appear, some broad band noise will simultaneously appear if there is a strange attractor. Practically, the three torus can decay (into a strange attractor) immediately after the ciritical parameter value for its existence has been reached, such that one observes in the power spectrum only two independent frequencies, that is, two Hopf bifurcations and then chaos as shown in Fig. 96.

It is understandable that chaotic motion only becomes possible after two Hopf bifurcation, when the trajectory can explore additional dimensions, because doubly periodic motion corresponds to a trajectory on a torus (i. e. on a two-dimensional manifold), on which chaos is forbidden by the Poincaré-Bendixson theorem. However, Ruelle and Takens (1978) showed that a strange attractor is not only possible, but that there exist certain perturbations which definitely convert quasi-periodic motion on a three torus into chaotic motion on a strange attractor. The subtle point is that these perturbations can be infinitesimal; however, not all infinitesimal perturbations will lead to a destruction of the three torus such that the probability for the appearance of a strange attractor nevertheless can be small. The resulting attractor is, in contrast to the three-torus, robust with respect to small changes in the paramters of the system.

The proof of this theorem is mathematically too involved to be presented here. Instead, we will proceed as follows. First we will present the results of a numerical experiment by Grebogi, Ott and Yorke (1983) who investigated the strength of the perturbation which is needed to destroy a three torus in favor of a strange attractor. Basically, they confirmed numerically the theorem of Ruelle Takens and Newhouse (1978), which had been obtained analytically, but their calculations suggest that *smooth perturbations must have a finite strength* in order to generate a strange attractor from a three torus. Next, we describe two experimental examples where three independent frequencies have been observed together with broadband noise in the power spectrum, and two experiments in which the destruction of a two torus into a strange attractor has been observed by reconstruction of the Poincaré map. Since, in the last two experiments, the third frequency has not been observed, one could ask whether there exists a direct transition from a two torus to a strange attractor that is independent of the Ruelle Takens Newhouse (1978) mechanism. We leave this question open, because such a transition could be interpreted as an example of the Ruelle Takens Newhouse scenario in which the three torus becomes destroyed by an infinitesimal perturbation at the very moment as it is about to appear, so that one practically observes the decay of a two torus into a strange attractor (see however Curry and Yorke, 1978 for another interpretation). Finally, we discuss in the next section, universal features of the transition from (two frequency) quasiperiodicity to chaos in terms of simple circle maps and describe some relevant experiments.

Possibility of Three-Frequency Quasiperiodic Orbits

Newhouse, Ruelle and Takens (1978) showed that, in a system with a phase-space flow consisting of three incommensurate frequencies, arbitrarily small changes to the system convert the flow from a quasiperiodic three-frequency flow to chaotic flow.

One might naively conclude that three-frequency flow is improbable since it can be destroyed by small perturbations. However, it has been shown numerically by Grebogi, Ott and Yorke (1983) that the *addition of smooth nonlinar perturbations does not typically destroy three-frequency quasiperiodicity.* (In the proof by Newhouse et al., the small perturbations required to create chaotic attractors have small first and second derivatives, but do not necessarily have small third- and higher-order derivatives, as expected for physical applications.)

The calculation by Grebogi et al. (1983) can be summarized as follows: According to Section 6.2, the Poincaré map associated with a flow having two incommensurate frequencies (perturbed by $\varepsilon f(\theta)$) can be described by the map (6.13):

$$\theta_{n+1} = \theta_n + \Omega + \varepsilon f(\theta_n) \tag{6.6.}$$

where $f(\theta)$ is periodic in θ, and θ_n is taken modulo 1. By analogy, a flow with three incommensurate frequencies corresponds to a map:

$$\theta_{n+1} = \theta_n + \omega_1 + \varepsilon P_1(\theta_n, \varphi_n) \tag{6.7a}$$

$$\varphi_{n+1} = \varphi_n + \omega_2 + \varepsilon P_2(\theta_n, \varphi_n) \tag{6.7b}$$

where θ_n and φ_n are again taken modulo 1, and $P_{1,2}$ are periodic in θ_n and φ_n. The parameters ω_1 and ω_2 are incommensurate with each other and with unity; that is, integers p, r, q do not exist for which $p\omega_1 + q\omega_2 + r = 0$.

By expressing $P_{1,2}$ as a Fourier sum of terms

$$A_{r,s} \sin [2\pi(r\theta + s\varphi + B_{r,s})] \tag{6.8}$$

Table 10: Observed frequencies for different types of attractors.

Type of attractor	Liapunov exponents	$\dfrac{\varepsilon}{\varepsilon_c} = \dfrac{3}{8}$	$\dfrac{\varepsilon}{c} = \dfrac{3}{4}$	$\dfrac{\varepsilon}{\varepsilon_c} = \dfrac{9}{8}$
Three-frequency quasiperiodic	$\lambda_1 = \lambda_2 = 0$	82%	44%	0%
Two-frequency quasiperiodic	$\lambda_1 = 0 \; \lambda_2 < 0$	16%	38%	33%
Periodic	$\lambda_1 < 0 \; \lambda_2 < 0$	2%	11%	31%
Chaotic	$\lambda_1 > 0$	0%	7%	36%

and retaining (somewhat arbitrarily) only the terms $(r, s) = (0, 1), (1, 0), (1, 1) (1, -1)$, Grebogi et al. calculated the Liapunov exponents λ_1, λ_2 for the map (6.7) for random values of $\omega_1, \omega_2, A_{r,s}$, and $B_{r,s}$. Their results are summarized in Table 11, which shows that for a fixed typical choice of $P_{1,2}$, the measure of (ω_1, ω_2) yielding chaos approaches zero as $\varepsilon \to 0$. Three-frequency quasiperiodicity is possible only when $\varepsilon < \varepsilon_c$ where the map is invertible.

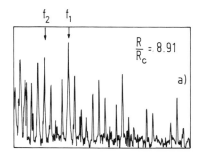

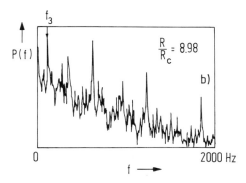

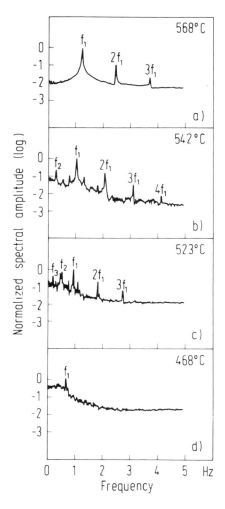

Fig. 98: Log-linear plot of the power spectrum (of the local temperature) in a Bénard experiment with mercury in a magnetic field. a) Quasiperiodic region with two incomensurate frequencies f_1 and f_2; b) three-frequency periodicity, i. e. f_1, f_2 and f_3 are present together with self-generated noise which decays exponentially. (Libchaber et al., 1983.)

Fig. 99: Power spectrum of the voltage across a BSN crystal through which a constant dc-current is maintained. With decreasing temperature, one observes a transition from one→two→three fundamental frequencies to chaos (after Martin, Leber and Martienssen, 1984).

The data in this table were computed using 256 random values of (ω_1, ω_2). The Liapunov exponents have been determined to the order 10^{-4} (Grebogi et al., 1983a).

A transition from quasiperiodicity to chaos which still exhibits three-frequency quasiperiodicity (i.e. the decay of this state to a strange attractor is not complete) has been observed by Libchaber, Fauve, and Laroche (1983) in a Bénard experiment with mercury in a magnetic field (see Fig. 98) and by Martin, Leber and Martienssen (1984) in the voltage spectrum of a ferroelectric Barium-Sodium-Niobate (BSN) crystal (see Fig. 99).

In the first case, the horizontal field serves as a second control parameter and additionally increases the viscosity of the electrically conducting fluid.

In the second case, the $Ba_2 Na Nb_5 O_{15}$ crystal, which displays a nonlinear current-voltage characteristic, is placed into a heating oven through which a constant flow of humidified oxygen is maintained (part of the conduction mechanism is due to oxygen vacancies). A stabilized dc-current is applied along the c-axis of the sample and one measures the voltage across the crystal together with the birefringence pattern. With increasing voltage, "domains" emerge from the cathode and disperse gradually through the crystal (see Plate IV at the beginning of the book). Since there are *three* control parameters (temperature, current density and oxygen flow), BSN provides an interesting system for experimental studies of chaos.

Break up of a Two Torus

It has been mentioned above that the conversion of quasiperiodic motion into chaotic motion on a strange attractor could occur apparently from a two torus if the three torus is so unstable that the third incommensurate frequency cannot be observed. Such transitions belong in principle also to the Ruelle-Takens-Newhouse scenario (see however Curry and Yorke, 1978) and have been seen in two hydrodynamic experiments.

Dubois and Bergé (1982) observed experimentally the emergence of a strange attractor in a *Bénard experiment*. They measured the time series of temperature $T(t)$ and reconstructed a two-dimensional Poincaré section by plotting $[T(t), \dot{T}(t)]$ at intervals $t = n\tau$, where $\omega_0 = 2\pi/\tau$ was determined from an independent measurement of the velocity. (This is another method of reconstructing an attractor from the measurement of one variable, note that in our example from chapter 5.3 $\vec{x}(t) = [\sin(2\pi t), \cos(2\pi t)]$ (eq. 5.43) the y component $y = \cos(2\pi t)$ could be obtained by differentiation i.e. $y \propto \dot{x}$). Fig. 100 shows how the Poincaré section, which consists of a closed loop (as expected for a section of a torus), develops into a strange attractor.

Another example for the emergence of chaos after two Hopf bifurcations has been observed after a Taylor instability by Swinney and Gollub (1978). The Taylor instability occurs in a fluid layer between an inner cylinder rotating with an angular velocity Ω and a stationary outer cylinder (see Fig. 101 and Plate III at the beginning of the book). For small Ω, angular momentum fed to the inner cylinder is transported outside by viscosity (a). Above a critical angular velocity Ω_c, this state becomes unstable, and momentum is transported by annular convection cells (b). At still higher Ω's, periodic and multiply periodic oscillations of these cells occur which merge into chaos after two Hopf bifurcations.

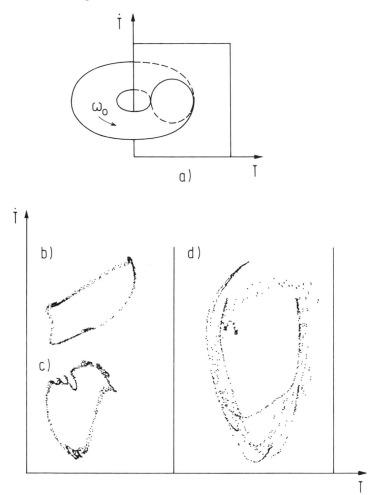

Fig. 100: Poincaré sections for the Bénard experiment: a) Schematic section through torus; b)–d) experiments showing with increasing Rayleigh number a transition form quasiperiodic motion (b) to substructures indication the destruction of the torus (c) and then to a strange attractor (d). (After Dubois, Bergé and Croquett, 1982.)

The following results in Fig. 102 have been obtained by reconstructing the phase space for a Taylor experiment from a time series of the radial velocity $\{v(t_k), \ldots, v(t_k + m\tau)\}$ with $t_k = k \cdot \tau_0$, $k = 0, 1, 2, \ldots, (\tau_0 < \tau)$:

a) The Poincaré section shows the break up of a torus similar to Fig. 100.

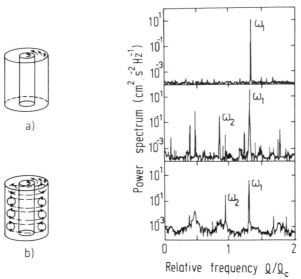

Fig. 101: The Taylor instability and power spectrum of the velocity (after Swinney and Gollub, 1978).

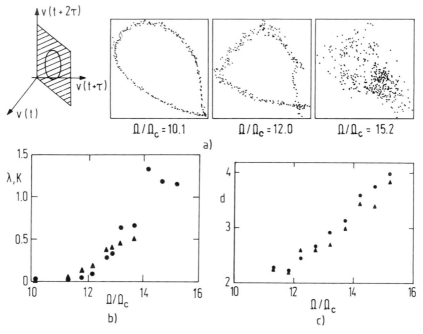

Fig. 102: Experimental properties of a strange attractor which occurs in a Taylor experiment: a) Plane of the Poincaré section and break up of the torus with increasing Ω. b) K-entropy (▲) and largest Liapunov exponent (●) vs. Ω/Ω_c. c) Hausdorff dimension D (●) and correlation dimension D_2 (▲) vs. Ω/Ω_c. (After Brandstäter et al., 1983.)

b) The *K*-entropy (obtained via eq. (5.109) and the largest Liapunov exponent λ (obtained from the separation of nearby orbits in five-dimensional phase space) become positive for $\Omega > \Omega^*$. This *proves experimentally* the existence of a *strange attractor*.

c) The Hausdorff dimension D (obtained via eq. (5.49) and D_2 (obtained via eq. (5.76) increase slowly with Ω/Ω_c. This shows that there are only a *few relevant degrees of freedom* eben at Ω-values that are 30% above the critical value $\Omega^* \approx 12\,\Omega_c$ at the onset of chaos.

6.2 Universal Properties of the Transition from Quasiperiodicity to Chaos

The transition from quasiperiodic motion on a two torus to chaotic motion has also investigated by studying simple maps (Feigenbaum and Kadanoff, 1982; Rand et al. 1982, 1983; Jensen et al., 1984).

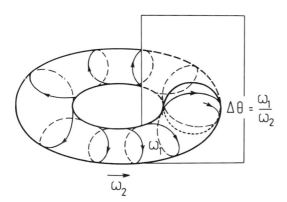

Fig. 103: Motion on a unit torus. For rational $\omega_1/\omega_2 = p/q$, the trajectory closes after q-cycles. This is called a *mode-locked* state. For irrational ω_1/ω_2, the motion is called *quasiperiodic;* the trajectory never closes and covers the whole torus.

Fig. 103 shows that the motion on an unperturbed unit torus can be described in polar coordinates by the Poincaré map

$$\theta_{n+1} = f(\theta_n) \equiv \theta_n + \Omega \quad \mathrm{mod}\,1 \,. \tag{6.9}$$

The parameter $\Omega = \omega_1/\omega_2$ determines the winding number

$$w = \lim_{n \to \infty} \frac{f^n(\theta_0) - \theta_0}{n} \tag{6.10}$$

which measures the average shift of the angle θ per iteration in eq. (6.10), the modulo in f has to be omitted. We find from eqns. (6.9–10) $w = \Omega$. But it should be noted that the definition of the winding number w given in eq. (6.10) holds for all maps of the unit circle onto itself.

In order to obtain an idea how eq. (6.9) should be modified to describe the break up of a torus in a physical system, we reconsider our kicked rotator from chapter 1, eq. (1.18), for the case that a constant torque $\Gamma\Omega$ has been added to the driving force. If we make, in eqns. (1.18 a, b), the following simplifying substitutions for $T = 1$:

$$x_n \to \theta_n ; \qquad \frac{e^\Gamma - 1}{\Gamma} y_n - \Omega \to r_n ; \qquad e^{-\Gamma} = b \tag{6.11 a}$$

$$Kf(\theta_n) \to \frac{\Gamma}{1 - e^{-\Gamma}} \, \frac{K}{2\pi} \sin (2\pi\theta_n) + \Gamma\Omega \tag{6.11 b}$$

we obtain

$$\theta_{n+1} = \theta_n + \Omega - \frac{K}{2\pi} \sin (2\pi\theta_n) + br_n \quad \text{mod } 1 \tag{6.12 a}$$

$$r_{n+1} = br_n - \frac{K}{2\pi} \sin (2\pi\theta_n) \tag{6.12 b}$$

where θ_n is the angle of the kicked rotator at time n, $r_n = y_n(e^\Gamma - 1)/\Gamma - \Omega$ is – apart from a constant shift – proportional to the angular velocity $y_n = \dot{\theta}(t = n)$. Eqns. (6.12 a–b) define the so called *dissipative circle map*. For vanishing nonlinearity ($K = 0$) and finite damping rate $b = e^{-\Gamma} < 1$, eqns. (6.12 a–b) reduce to the unperturbed map eq. (6.9) where Ω sets the rate of rotation.

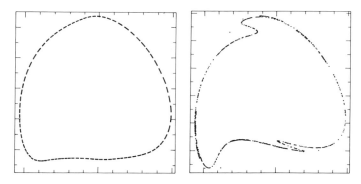

Fig. 104: Break up of a torus as described by the dissipative circle map (eqs. 6.12a, b) (after Bauer, priv. comm.).

Fig. 104 shows that the dissipative circle map indeed describes the break up of a torus if the parameter K which measures the strength of the nonlinearity $\sin (2\pi\theta_n)$ is increased from $K = 0.814$ to $K = 1.2$. In both pictures, we plotted $y_n = (1 + 4r_n)$ $\sin \theta_n$ versus $x_n = (1 + 4r_n) \cos \theta_n$ with θ_n and r_n from eq. (6.12) and $\Omega = 0.612$, $b = 0.5$.

These pictures should be compared to Figs. 100 and 102 which show the destruction of a torus in experimentally measured Poincaré maps. For strongly dissipative systems ($b \to 0$), the radial motion of the trajectory disappears in eqns. (6.12a, b), and they reduce to the *one dimensional circle map:*

$$\theta_{n+1} = f(\theta_n) \equiv \theta_n + \Omega - \frac{K}{2\pi} \sin(2\pi\theta_n) \quad \text{mod} \, 1 \,, \tag{6.13}$$

which describes the transition from quasiperiodicity to chaos only by the motion of the angles θ_n. *Here θ_n is again understood modulo 1;* K provides, in analogy to the Reynold's number, a measure for the nonlinearity $\sin(2\pi\theta_n)$ (which must be added to obtain a transition to chaos), and Ω sets again the rate of rotation (see eq. (6.10)). In the following section, we study the break up of the torus into a strange attractor via this map.

It will be shown below that (by analogy to the logistic map for the period-doubling route) the special form of $f(\theta)$ is rather unimportant, and, of more importance, are the following general features of $f(\theta)$:

– $f(\theta)$ has the property $f(\theta + 1) = 1 + f(\theta)$.

– For $|K| < 1$, $f(\theta)$ (and its inverse) exists and is differentiable (i. e. $f(\theta)$ is a diffeomorphism).

– At $K = 1$, $f^{-1}(\theta)$ becomes nondifferentiable, and for $|K| > 1$, no unique inverse to $f(\theta)$ exists.

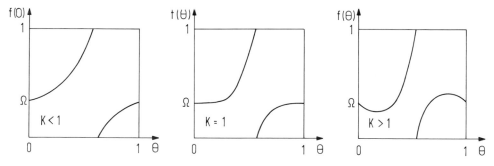

Fig. 105: Variation of the map $f(\theta)$ with the parameter K. Note, that for $K > 1$, the map becomes noninvertible.

To obtain an overview of the behavior of the circle map (6.13), we show in Plate XVI (at the beginning of this book) its Liapunov exponent λ depicted in colors as a function of the two control parameters K and Ω. We distinguish three regimes:

– For $|K| < 1$, one finds the so-called Arnold's tongues (Arnold, 1965) where the motion is mode locked; that is, the winding number w (see eq. (6.10)) is rational.

Between these tongues, the winding number is irrational. Both areas in the $K - \Omega$ plane, the mode locking and the nonmode locking one, are finite (see Fig. 106).

– At $K = 1$, the Arnold's tongues moved together in such a way that the nonmode locked Ω intervals from a self similar Cantor set with zero measure.

– For $|K| > 1$, the map becomes noninvertible, chaotic behavior becomes possible, but chaotic and nonchaotic regions are *densely* interwoven in parameter space (i. e. the $K - \Omega$ plane).

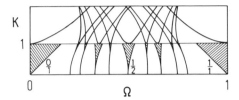

Fig. 106: Phase diagram of the circle map (schematically). $K < 1$: Within the Arnold tongues (hatched) the winding number w is rational, and one has mode locking. $K = 1$; the Arnold tongues moved together, the remaining nonmode-locked "holes" form a Cantor set. $K > 1$: Chaos becomes possible, but coexists with order. The lines correspond to the parameter values for superstable, that is nonchaotic cycles which are associated with the mode locking regions.

In the following section, we will investigate these different regimes in more detail. In the first part of this section, we study the nonchaotic mode-locking behavior. Mode locking means, according to eq. (6.10), that the ratio between the number of cycles, which the system executes divided by the number of oscillations of the driving force (think of a kicked rotator), is a rational number. Thus, mode locking with winding number $w = 1$ corresponds to complete synchronisation between the external force and the system. Since this phenomenon occurs very often in nature – already in the 17th century the Dutch physicist Ch. Huyghens observed synchronisation between two clocks hanging back-to-back on a wall – the understanding of mode locking in nonlinear systems is of considerable interest.

In the second part of this section, we investigate universal properties at the transition from quasiperiodicity to chaos using different renormalization group formalisms. Since one has two control parameters K and Ω one has to distinguish between *local* scaling behavior, which occurs *near a point* in the $K - \Omega$ plane, and *global* scaling behavior, which occurs *for a whole set of Ω values* and describes the merging together of the Arnold tongues as the line $K = 1$ is approached in Fig. 106.

It will be shown that the local transition from quasiperiodicity to chaos near an irrational winding number displays, as a function of the control parameter Ω in its renormalization group description, some formal analogies to the period doubling route. In contrast, the numerically found global scaling requires a different normalization group approach, and we will only calculate the universal Hausdorff dimension of the Cantor set which is formed by the nonmode locked Ω intervals at $K = 1$ (see Fig. 106).

Mode Locking and the Farey Tree

In this subsection, we investigate the mode locking which occurs in the interates of the circle map. It will be shown that for fixed K the width of an Arnold tongue decreases if the denominator q in the corresponding rational winding number $w = p/q$ increases. The resulting hierarchy of tongues at $K = 1$ can be conveniently represented by a Farey tree which orders all rationals in [0, 1] according to their increasing denominators (see Hardy and Wright, 1938).

For a general mode locked state with $w = p/q$, the corresponding Ω interval $\Omega = \Omega(K)$ can be calculated from the condition that a q-cycle with elements $\theta_1^* \ldots \theta_q^*$ occurs in the circle map (6.13):

$$f_{\Omega,K}^q(\theta_i^*) = p + \theta_i^* \tag{6.14}$$

which is stable i.e.

$$f_{\Omega,K}^{q\prime}(\theta_i^*) \mid = \mid \prod_{i=1}^q f_{\Omega,K}^{\prime}(\theta_i^*) \mid = \mid \prod_{i=1}^q [1\text{-}K \cos(2\pi\theta_i^*)] \mid < 1 . \tag{6.15}$$

(Here the indices K and Ω indicate that the left hand side in eqns. (6.14, 6.15) is still a function of both variables.)

Eqns. (6.13–15) yield, e.g., for $w = 1$:

$$f_{\Omega,K}(\theta_0) = \theta_0 \rightarrow \Omega = \frac{K}{2\pi} \sin(2\pi\theta_0) \tag{6.16}$$

and

$$|f_{\Omega,K}^{\prime}(\theta_0)| = |1 - K\cos(2\pi\theta_0)| < 1 . \tag{6.17}$$

For $|K| < 1$, the boundaries $|f_{\Omega,K}^{\prime}(\theta_0)| = 1$ are reached for $\theta_0 = \pm\pi/4$ which implies, via eq. (6.16), that the first Arnold tongue is a triangle with a width Ω

$$\Omega = \pm \frac{K}{2\pi} \tag{6.17}$$

as shown in Fig. 106.

The general eqns. (6.13–15) have been solved numerically by P. Bak and T. Bohr (1984) who found that for $0 < K < 1$ a whole interval $\Delta\Omega(p/q, K)$ of Ω values is associated to every rational winding number. For $K = 1$, these intervals form a complete self similar devil's staircase as shown in Fig. 107. The staircase for $K = 1$ is termed complete because the sum S of all Ω intervals is equal to 1 i.e.

$$S = \sum_{p,q} \Delta\Omega(p/q, 1) = 1 . \tag{6.18}$$

For $0 < K < 1$, the staircase becomes incomplete i.e. $S < 1$.

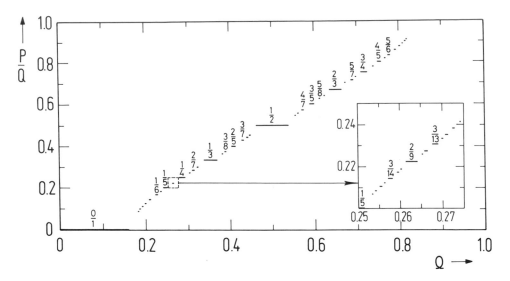

Fig. 107: The mode locking structure of the circle map, eq. (6.13) at $K = 1$. The devil's staircase is complete; the numbers denote the rational winding numbers (after Jensen at al., 1984).

Fig. 107 shows that the widths of the steps becomes smaller if the denominator in the corresponding winding number increases. Furthermore, if we have two steps with winding numbers p/q and p'/q', then the largest step in between has a winding number $(p + p')/(q + q')$. If we list a few examples: $0/1 < 1/2 < 1/1$; $1/2 < 2/3 < 1/1$; $1/2 < 3/5 < 2/3$, etc., we see that $(p + p')/(q + q')$ is the rational number with the smallest denominator which lies between p/q and p'/q'. Thus the Farey tree, shown in Fig. 108, which orders all rationals p/q in [0, 1] with increasing denominators q, orders

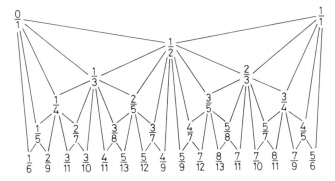

Fig. 108: The Farey tree orders all rationals in [0, 1] with increasing denominators according to the rule that the largest rational between p/q and p'/q' is $(p + p')/(q + q')$ (after Cvitanović and Söderberg, 1985 a).

also all mode locking steps with $w = p/q$ in the circle map according to their decreasing widths.

Up to now, our observations were only based on numerical evidence shown in Fig. 107; however, there exists also a simple analytical result that establishes the relation between the devil's staircase and the Farey tree.

The monotony of the circle map (and its iterates) in Ω implies that to every winding number in the Farey tree belongs exactly one mode locking step in the devil's staircase. Suppose one has a superstable q-cycle $f^q_{\Omega(p,q)}(\theta) = p + \theta$ and a q'-cycle $f^{q'}_{\Omega(p',q')}(\theta')$. If we combine both iterations we obtain:

$$f^q_{\Omega(p,q)}[f^{q'}_{\Omega(p',q')}(\theta)] = p + p' + \theta \tag{6.19}$$

that is, a cycle with the winding number $(q + q')/(p + p')$. Increasing $\Omega(p, q)$ in $f^q_{\Omega(p,q)}$ overshoots this cycle. This can be compensated by reducing $\Omega(p', q')$ in $f^{q'}_{\Omega(p',q')}$. Due to the fact that both iterates are monotonous in Ω, one can repeat this procedure until both Ω values coincide. Hence, the Ω interval between the p/q and p'/q' cycles, always contains an Ω value which corresponds to a $(p + q')/(q + q')$ cycle as claimed above.

The Farey tree construction has a universal importance because it orders not only the mode locking regions for the circle map but also for real systems such as a driven pendulum, Josephson junctions, and sliding charge density waves. This of course means that the dynamics of these systems can be reduced to circle maps as will be shown in sect 6.3.

Local Universality

The transition from quasiperiodicity to chaos is characterized by two types of universality. One is associated with the transition from quasiperiodicity to chaos for a special, that is, local winding number, and it shows close parallels to the period doubling route. Its experimental verification is difficult because minute changes in winding numbers lead to large changes in scaling behavior. The second type is called global universality and pertains to a whole range of winding numbers. It describes the scaling behavior of the set of Ω values, complementary to the Arnold tongues on which the dynamical system is mode locked, and it has been observed experimentally in several systems.

We begin with an investigation of the transition from quasiperiodicity to chaos for the golden mean winding number, because it also forms the basis for the investigation of the global universality for the circle map.

In order to observe a transition from quasiperiodicity to chaos in the iterates of (6.13), *two* parameters have to be adjusted. If we increase, for example, the nonlinearity via K, Ω must alsways be balanced to keep the winding number w fixed to a given irrational values (this guarantees quasiperiodicity). But how can this be performed for a winding number which still gives the average shift of θ per iteration, and which, however, for general maps has to be defined as the limit (see eq. 6.10):

$$w = \lim_{n \to \infty} \frac{f^n(\theta_0) - \theta_0}{n} \tag{6.20}$$

(where the modulo in f has to be omitted)? We use the following method which has been suggested by Greene (1979) (in a similar context for Hamiltonian systems). One calculates for fixed K the value $\Omega_{p,q}(K)$ which a) belongs to a q-cycle of the map $f(\theta)$, b) contains $\theta = 0$ as an element, and c) provides a shift by p. Thus, $\Omega_{p,q}$, which generates a rational winding number $w = p/q$, is defined by

$$f^q_{K,\Omega}(0) = p . \tag{6.21}$$

Next, the irrational winding number is approximated by a sequence of truncated continued fractions, i.e. rationals. If we consider, for example, the winding number $w^* = (\sqrt{5} - 1)/2$ which has as a continued fraction of the simple form

$$w^* = \cfrac{1}{1 + \cfrac{1}{\dots}} \tag{6.22}$$

then the so-called *Fibonacci numbers* F_n, which are defined by

$$F_{n+1} = F_n + F_{n-1} ; \quad F_0 = 0 , \quad F_1 = 1 ; \quad n = 0,1,2,\dots \tag{6.23}$$

via

$$w_n \equiv \frac{F_n}{F_{n+1}} = \frac{F_n}{F_n + F_{n-1}} = \tag{6.24a}$$

$$= \cfrac{1}{1 + \cfrac{F_{n-1}}{F_n}} = \underbrace{\cfrac{1}{1 + \cfrac{1}{1 + \dots}}}_{n \text{ times}} \tag{6.24b}$$

yield a sequence of rationals w_n which converges towards

$$w^* = \lim_{n \to \infty} w_n . \tag{6.25}$$

For $n \to \infty$ eqs. (6.24a, b) yield

$$w^* = \cfrac{1}{1 + w^*} \to w^{*2} + w^* - 1 = 0 \to w^* = (\sqrt{5} - 1)/2 . \tag{6.27}$$

This number is the so-called *golden mean*, which is defined in geometry by sectioning a straight line segment in such a way that the ratio of the longer segment *l* to the total length *L* equals the ratio of the shorter segment to the longer segment, i. e. $w^* = l/L = (L - l)/l$. In the following, we confine ourselves to this special winding number $w^* = (\sqrt{5} - 1)/2 = 0.6180339$, which is the "worst" irrational number in the sense that it is least well approximated by irrationals (see eqs. (6.22–24)). Although any given irrational number has a unique representation by continued fractions, the renormalization scheme has, up to now, only been applied to the so-called quadratic irrationals, which are the solutions of a quadratic equation with integer coefficients, and for which the continued fraction representation is periodic.

Using the procedure described above, Shenker (1982) obtained the following *numerical results* for the circle map (6.13):

a) The values of the parameters $\Omega_n(K)$ (6.13) which via (6.21) generate the winding numbers w_n in (6.24), geometrically tend to a constant, i. e.

$$\Omega_n(K) = \Omega_\infty(K) - \text{const} \cdot \tilde{\delta}^{-n} \tag{6.28a}$$

where

$$\tilde{\delta} = \begin{cases} -2.6180339\ldots = -w^{*-2} & \text{for} \quad |K| < 1 \\ -2.83362\ldots & \text{for} \quad |K| = 1 \end{cases} \tag{6.28b}$$

is a universal constant that, however, depends on w^*.

b) The distances d_n from $\theta = 0$ to the nearest element of a cycle which belongs to w_n

$$d_n = f_{\Omega_n}^{F_n}(0) - F_{n-1} \tag{6.29a}$$

scale like

$$\lim_{n \to \infty} \frac{d_n}{d_{n+1}} = \tilde{a} \tag{6.29b}$$

where $\tilde{a}$ is again a universal constant with values

$$\tilde{a} = \begin{cases} -1.618\ldots = -w^{*-1} & \text{for} \quad |K| < 1 \\ -1.28857\ldots & \text{for} \quad |K| = 1 \end{cases} \tag{6.29c}$$

(Note, that $f_{\Omega_n}^{F_{n+1}}(0) - F_n = 0$).

c) Fig. 109 shows the periodic function

$$u(t_j) = \theta^n(t_j) - t_j; \quad j = 0, 1, 2, \ldots \tag{6.30}$$

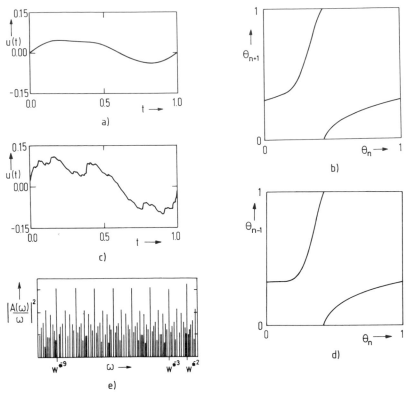

Fig. 109: a) $u(t)$ for $K = 0.5$ (after Shenker, 1982);) b) the map (5.57) at $K = 0.5$ and $w = w^*$ is a diffeomorphism (after Jensen et al. 1983 a); c) $u(t)$ becomes bumpy for $K = 1$ (after Shenker, 1982); d) the map (5.57) at $K = 1$ and $w = w^*$ becomes noninvertible $f'(0) = 0$ (after Jensen et al., 1983 a); e) the power spectrum for $K = 1$ (after Rand et al., 1983). Note that indeed for $n \to \infty$, $u(t_j)$ converges to a function in $0 \le t \le 1$ because the periodicity of u implies that its argument is taken mod 1 and $j \cdot w^*$ mod 1, $j = 0, 1, 2, \ldots n$ covers [0, 1].

that measures the time dependence of the cycle elements

$$\theta^n(t_j) \equiv \theta(j \cdot w_n) \equiv f^j(0) \tag{6.31}$$

for times $t_j \equiv j \cdot w_n$ in the limit $n \to \infty$.

(Here, $f^j(0)$ is taken at $\Omega_n(K)$, and $u(t_j)$ is periodic since the property $f(\theta + 1) = f(\theta) + 1$ leads to $\theta(t_j + 1) = \theta(t_j) + 1$). For $|K| < 1$ and $\Omega_n \to \Omega_\infty$, the variable $u(t)$ varies smoothly with t, but its behavior becomes "bumpy" for $|K| = 1$, which signals the transition from quasiperiodicity to chaos.

d) The power spectrum

$$A(\omega) = \frac{1}{F_{n+1}} \sum_{j=0}^{F_{n+1}-1} u(t_j) e^{2\pi i \omega t_j} \tag{6.32}$$

Table 11: Parallels between the Transitions to Chaos via Period Doubling and Quasiperiodicity.

Period Doubling	Quasiperiodicity
Logistic map	Circle map
$x_{n+1} = f_r(x_n) \equiv r_{x_n}(1 - x_n)$	$\theta_{n+1} = f_{K\Omega}(\theta_n) \equiv \theta_n + \Omega - \dfrac{K}{2\pi}\sin(2\pi\theta_n)\,\mathrm{mod}\ 1$
One control parameter r	Two control parameters $K,\ \Omega$
At $r = R_n$ superstable cycle of length 2^n	At $\Omega = \Omega_n$ superstable cycle of length F_{n+1}

R_n is calculated from	Ω_n is calculated from
$f_{R_n}^{2^n}(0) = 0$ (cycle closes)	$f_{K,\Omega_n}^{F_{n+1}}(0) - F_n = 0 \left(\text{ensures } w_n = \dfrac{F_n}{F_{n+1}}\right)$
Parameter scaling	
$R_{n+1} - R_n \sim \delta^{-n}$ for $n \gg 1$	$\Omega_{n+1} - \Omega_n \sim \tilde{\delta}^{-n}$ for $n \gg 1$
Scaling of distances between cycle elements	
$d_n \equiv f_{R_n}^{2^n}(0)$	$d_n \equiv f_{K,\Omega_n}^{F_n}(0) - F_{n-1}$
(compare to $f_{R_n}^{2^{n-1}}(0) = 0$)	(compare to $f_{K,\Omega_n}^{F_{n+1}}(0) - F_n = 0$)
$\dfrac{d_n}{d_{n+1}} = -\alpha$ for $n \gg 1$	$\dfrac{d_n}{d_{n+1}} = \tilde{\alpha}$ for $n \gg 1$

for $\omega = 0, \ldots F_{n+1}$ is shown for $n \to \infty$ in Fig. 109e. It displays self-similarity (the major structure between any two adjacent peaks is essentially the same), and the main peaks occur at powers of the Fibonacci numbers reflecting the fact that the motion is almost periodic after F_n iterations.

These results (especially a) and b)) appear very similar to those found for the perioddoubling route, and it is therefore natural to attempt a *renormalization-group treatment* of this transition which establishes its universal features. The formal parallels between the transitions to chaos via period doubling and quasiperiodicity are summarized in table 11. (Note, that $\tilde{a}$ and $\tilde{\delta}$ in eqs. (6.28–29) are different from the Feigenbaum constants.)

To derive the corresponding functional equations, we define (see eq. (6.29b)) the functions

$$f_n(x) \equiv \tilde{a}^n f^n(\tilde{a}^{-n}x) \qquad \text{where} \tag{6.33}$$

$$f^n(x) \equiv f^{F_{n+1}}(x) - F_n \tag{6.34}$$

such that eq. (6.29b) becomes

$$\lim_{n \to \infty} \tilde{a}^n d_n \propto \lim_{n \to \infty} \tilde{a}^n f^n(0) = \lim_{n \to \infty} f_n(0) = \text{const} . \tag{6.35}$$

As in the case of period doubling (see eq. 3.15), this relation indicates that the sequence $\{f_n(x)\}$ converges towards a universal function

$$\lim_{n \to \infty} f_n(x) = f^*(x) \tag{6.36}$$

where $f^*(x)$ is again the solution of a fixed-point equation which we shall now derive. More precisely, we consider f_n at $\Omega = \Omega_\infty$, which corresponds to $i \to \infty$ in eq. (3.21).

The function f^{n+1} can be obtained from f^n and f^{n-1} by a rule which is dictated by the recursion of the Fibonacci numbers (6.23) and the property $f(x + 1) = f(x) + 1$:

$$\begin{aligned} f^{n+1}(x) &= f^{F_{n+2}}(x) - F_{n+1} = \\ &= f^{F_{n+1}}[f^{F_n}(x)] - (F_{n+1} + F_n) = \\ &= f^n[f^{n-1}(x)] . \end{aligned} \tag{6.37}$$

Because the operation of iteration is commutative, we also have

$$f^{n+1}(x) = f^{n-1}[f^n(x)] . \tag{6.38}$$

According to eqs. (6.37–38), there are now *two* ways of calculating $f_{n+1}(x)$:

$$f_{n+1}(x) = \tilde{a}f_n[\tilde{a}f_{n-1}(\tilde{a}^{-2}x)] \tag{6.39a}$$

and

$$f_{n+1}(x) = \tilde{a}^2 f_{n-1}[\tilde{a}^{-1}f_n(\tilde{a}^{-1}(x)] . \tag{6.39b}$$

Both equations become equivalent for the initial conditions

$$f_0 [\tilde{a}^{-1} f_1 (\tilde{a} x)] = \tilde{a}^{-1} f_1 [\tilde{a} f_0 (x)] . \qquad (6.40)$$

Taking the limit $n \to \infty$ in (6.39a), we obtain for the fixed point function

$$f^* (x) = \tilde{a} f^* [\tilde{a} f^* (\tilde{a}^{-2} x)] . \qquad (6.41)$$

One can immediately verify that

$$\bar{f}^* (x) = -1 + x \qquad (6.42)$$

is a rigorous solution to this equation. If we substitute $\bar{f}^* (x)$ into (6.41), we obtain

$$-1 + x = -\tilde{a}^2 - \tilde{a} + x \to \tilde{a} = -w^{*-1} . \qquad (6.43)$$

This value for $\tilde{a}$ (which is equal to the second solution of eq. (6.27)) agrees with the numerical result for $|K| < 1$ (see eq. (6.29c)).

For $|K| = 1$, we expect that (6.41) has a different solution because the linear terms is then absent in our model equation (6.13):

$$f(0) = \Omega + \theta^3 \cdot \text{const.} , \quad \text{for} \quad \theta \to 0 ; \quad |K| = 1 . \qquad (6.44)$$

If for $|K| = 1$ we try the ansatz

$$f^* (x) = 1 + ax^3 + bx^6 ... \qquad (6.45)$$

a value for $\tilde{a}$ is found which is consistent with eq. (6.29c). This establishes the universality of the $\tilde{a}$'s for $|K| \le 1$.

By analogy to period doubling, the δ's appear as eigenvalues of the linearized fixedpoint equation. These equations are somewhat more complicated than in the Feigenbaum route because the recursion relations are of *second* order; that is, f_n *and* f_{n-1} are required to produce f_{n+1} (for more details see, e.g., the article of Feigenbaum, Kadanoff and Shenker, 1982).

Global Universality

Let us now consider the globally universal properties of the set of Ω values that is complementary to the Arnold tongues and corresponds to irrational winding numbers. The following numerical results have been obtained by Jensen, P. Bak and T. Bohr (1984):

- For $K \to 0$ (from below), the complement C of the total length of the steps in the (incomplete) devil's staircase, i.e. $C = 1 - \sum_{p,q} \Delta\Omega(p/q, K)$, decreases to zero with a power law

  $$C \propto (1 - K)^\beta$$

  where the exponent $\beta \cong 0.34$ is the same for all $f(\theta)$ in eq. (6.13) which have a cubic inflection point at $K = 1$.

- At $K = 1$, the Ω values belonging to irrational winding numbers form a self similar thin cantor set (of zero measure) whose Hausdorff dimension $D^* \equiv 0.87$ is again universal.

Whereas there exists up to now no theoretical explanation for the value of β, we will follow the work of Cvitanovich et al. (1985 b) and calculate D^* by introducing a whole family of universal functions that maintains a dependence on Ω (which was lost in the previous R. G. formulation where we put $\Omega = \Omega_\infty$ in eq. (6.36)).

For simplicity, we explain this method first for the period doubling route and transfer it then to the circle map. Fig. 110 shows again the self similar structure of the bifurcation tree from section 3.1.

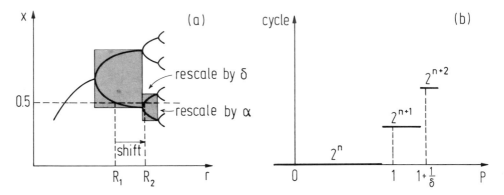

Fig. 110: a) Self similarity of the bifurcation tree; b) stability intervals of 2^n, 2^{n+1}, and 2^{n+2} cycles of $g_p(x)$ with n "arbitrarily high". (After Cvitanović, 1984.)

In order to capture the change in x and r, we follow the procedure of Cvitanović (1984) and introduce the modified doubling operator $\hat{T}$. It denotes the operation of iteration twice, rescaling x by a, shifting r to the corresponding values (with the same slope at the cycle points) at the next bifurcation, and rescaling it by δ:

$$\hat{T} f_{R_n + p\Delta_n}(x) \equiv -a f_{R_{n+1} + p\Delta_{n+1}}^{(n+1)}(-x/a) \tag{6.46}$$

$$= -a f_{R_n + \Delta_n(1 + p/\delta_n)}^{(n)} [f_{R_n + \Delta_n(1 + p/\delta_n)}^{(n)}(-x/a)]$$

where

$$f^{(n)}(x) = f^{2^n}(x), \quad \varDelta_n = R_{n+1} - R_n, \quad \delta_n = \varDelta_n/\varDelta_{n+1} \tag{6.47}$$

and $0 \leqslant p \leqslant 1$ is a parameter which interpolates the r's between a 2^n and a 2^{n+1} cycle. If we call $\lim\limits_{n\to\infty} \delta_n = \delta$ and

$$\lim_{n\to\infty} \hat{T}^n f_{R_0 + p\varDelta_0}(x) \equiv g_p(x), \tag{6.48}$$

we obtain from the definitions (6.46–48) an equation for the universal family of functions $g_p(x)$:

$$g_p(x) = \hat{T} g_p(x) = -a g_{1+p/\delta}[g_{1+p/\delta}(-x/a)] \tag{6.49}$$

with boundary conditions:

$$g_0(0) = 0 \quad \text{and} \quad g_1(0) = 1 . \tag{6.50}$$

The first condition means that the origin of p corresponds to the superstable fixed point. The second condition sets the scale of x and p by the superstable two cycle (see Fig. 111)

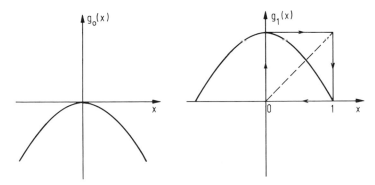

Fig. 111: Boundary conditions for the functions $g(x)$ and $g_1(x)$ defined in the text (after Cvitanović, 1984).

Note that our fixed point equation (3.22) for the usual doubling operator can be obtained from by choosing $p = p^*$ such that $g_{p^*} = g_{1+p^*/\delta}$ i.e. $p^* = \delta/(\delta - 1)$. The family of universal functions $g_p(x)$ in eq. (6.49) is called the unstable manifold (Vul and Khanin, 1982) because operation with $\hat{T}$ drives $g_p(x)$ away from the fixed point values $g_{p^*}(x)$.

The advantage of eq. (6.49) with respect to the usual fixed point equation (3.22) is twofold. First, one could obtain *both* Feigenbaum constants a and δ from eq. (6.49) by expanding $g_p(x)$ into a double power series in p and x and comparing the coefficients of equal powers, that is one needs no linearization around the fixed point.

Since we have calculated their values already in section 3.2, we now concentrate on the second aspect of eq. (6.49), namely its stability interpretation. It follows from Fig. 110b and eq. (6.49) that if $g_p(x)$ has a stable 2^n cycle (with n "arbitrarily high") in the interval-$p_0 < p < p_0$, then $g_{1+p/\delta}(x)$ has a stable 2^{n+1} cycle in the p-interval around 1 whose width is reduced by a factor of $1/\delta$.

This somewhat trivial looking statement becomes rather powerful if we translate eqns. (6.46–6.49) to the circle map where they will allow us to obtain some insight into the self-similarity of the width of the mode-locking regions. The period doubling operation (eq. (3.22)) translates according to eqns. (6.37–39) into:

$$\mathrm{T}\,[f^{n-1}, f^{n-2}] = \tilde{a}\, f^{n-1}\,[f^{n-2}\,(x/\tilde{a}^2]\qquad(6.51)$$

where $f^n(x) \equiv f^{F_{n+1}}(x) - F_n$. Accordingly, the doubling plus x, r rescaling transformation $\hat{\mathrm{T}}$ from eq. (6.46) changes into:

$$\tilde{\mathrm{T}}[f^{n-1}_{\Omega_{n-1}+p\Delta_{n-1}}(x), f^{n-2}_{\Omega_{n-2}+p\Delta_{n-2}}(x)] \equiv \tilde{a}\, f^{n-1}_{\Omega_n+p\Delta_n}\,[\tilde{a}\, f^{n-2}_{\Omega_n+p\Delta_n}\,(x/\tilde{a}^2)]\qquad(6.52)$$

$$= \tilde{a}\, f^{n-1}_{\Omega_{n-1}+(1+p/\delta_n)\Delta_{n-1}}\,[\tilde{a}\, f^{n-2}_{\Omega_{n-2}+(1+p/\delta_{n-1}+p/\delta_{n-1}\delta_n)\Delta_{n-2}}\,(\tilde{x}/a^2)]$$

where

$$\Delta_n = \Omega_{n-1} - \Omega_n\,;\; \tilde{\delta}_n = \Delta_n/\Delta_n + 1$$

and $0 \le p \le 1$ interpolates the Ω's between a F_n and a F_{n+1} cycle. Calling

$$\lim_{n\to\infty} \tilde{\mathrm{T}}^n\,[f^{n-1}_{\Omega_0+p\Delta_0}, f^{n-2}_{\Omega_{-1}+p\Delta_{-1}}] = \tilde{g}_p(x)\qquad(6.53)$$

eqns. (6.52 – 53) yield (in analogy to eq. (6.49):

$$\tilde{g}_p(x) = \tilde{a}\,\tilde{g}_{1+p/\tilde{\delta}}\,[a\tilde{g}_{1+p/\tilde{\delta}+p/\tilde{\delta}^2}\,(x/\tilde{a}^2)]\qquad(6.54)$$

where the normalization conditions are again $\tilde{g}_0(0) = 0$, $\tilde{g}_1(0) = 1$.

One could again determine from eq. (6.54) the parameters $\tilde{a}$ and $\tilde{\delta}$ for the route from quasiperiodicity to chaos. But we will use here the universal object $\tilde{g}_p(x)$ to investigate the structure of mode lockings. For $p = 0$, $\tilde{g}_p(x)$ has, by construction, a superstable fixed point at $x = 0$. Our arguments are now closely parallel to those which we used to interpret eq. (6.49) in connection with Fig. (106). The range of p around zero, for which $\tilde{g}_p(x)$ still has a fixed point, is the range of parameters for which the original map is locked (in some winding number w_n with "infinitely large" n (see Fig. 112). However, around $p = 1$, there is another locked state which corresponds to the next locked region in the sequence, and the width of this region is scaled down by a factor of $1/\tilde{\delta}$ compared to the first (note $\tilde{\delta} < 0$ for the transition form quasiperiodicity to chaos). Around $p = 1 + 1/\tilde{\delta}$, there is another mode locked region scaled down by $1/\tilde{\delta}^2$ compared to the first, etc. Thus, by studying the stability of the fixed point of $\tilde{g}_p(x)$, one can find an infinity of mode locked states which are universally located. Although these are not all mode locked regions (see Cvitanović et al., 1985 b), they are sufficient to yield an estimate for the Hausdorff Dimension $D*$ of the "holes" in the devil's staircase at $K = 1$.

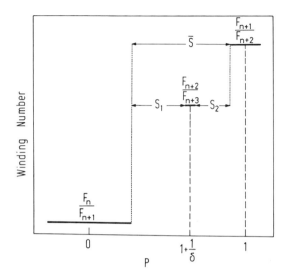

Fig. 112: Three steps in the universal devil's staircase as generated by the stability intervals of $\tilde{g}_p(x)$ and eq. (6.54). The widths of the steps are $p*$, $p*/\tilde{\delta}$ and $p*/\tilde{\delta}^2$ for F_n/F_{n+1}, F_{n+1}/F_{n+2}, F_{n+2}/F_{n+3}, respectively, where n is "arbitrarily high". Also indicated are the widths of the "holes" used in eq. (6.55) (after Cvitanović et al., 1985).

Generally, the Hausdorff Dimension of a self similar fractal can be computed from the equation (Hentschel and Procaccia, 1983):

$$\sum_i \left(\frac{s_i}{\bar{s}}\right)^D = 1 \tag{6.55}$$

where $\bar{s}$ is the length of a box that covers the whole set of points and s_i are the linear dimension of smaller boxes that also provide a complete coverage. Eq. (6.55) can be derived by noting that the number N of points of a fractal, which can be partitioned into boxes of size s_i containing N_i points, can be written as

$$N = \sum_i N_i . \tag{6.57}$$

Dividing this by N and using $(N_i/N) = (s_i/\bar{s})^D$ (which generalizes the formula that the number of points in a ordinary cube of linear dimension l grows like l^3), one obtains eq. (6.55). The values s_i and $\bar{s}$, which are needed to compute $D*$, can be read from Fig. 112. The range of parameter values, for which $\tilde{g}_p(x)$ has a stable fixed point, follows from $\tilde{g}_{p*}(x^*) = x^*$ and $|\tilde{g}'_{p*}(x^*)| = 1$. Since $\tilde{g}_p(x)$ is universal, so is p^* and, therefore, D^*, which is computed from the universal quantities p^* and $\tilde{\delta}$. If we estimate p^* crudely by $p^* = 1/2\pi$ (which is just the width of the first Ω step in the circle map, see eq. (6.17)). We obtain from eq. (6.55) and (Fig. 112), the value $D^* \cong 0.92$ which is less than 10% off the numerical result $D^* = 0.87$ found by Jensen et al. (1983 b). More accurate theoretical values for D^* can be obtained by considering better approximations to $g_p(x)$ and p^* (see Cvitanović et al., 1985 b).

6.3 Experiments and Circle Maps

There exists a large variety of real systems (see below) whose dynamical behavior can be modeled by circle maps. Usually an analysis to detect circle map behavior proceeds as follows:

- The power spectrum of the (Fourier transformed) measured signal shows two or three incommensurate frequencies before the onset of broadband noise. This indicates a transition from quasiperiodicity to chaos.

- A reconstruction of the trajectory in phase space, from the measurement of a single variable, shows the destruction of a torus in favour of a strange attractor. By choosing a proper plane in phase space, the torus section appears (before the transition to chaos) as a closed curve which can be parametrized by θ_n. A plot of θ_{n+1} versus θ_n reveals the existence (or nonexistence) of a circle map $\theta_{n+1} = f(\theta_n)$ with $f(\theta + 1) = f(\theta) + 1$.

- An analysis of the time series of the measured angles θ_n as a function of the experimental control of parameters reveals universal properties near the transition such as:
  - Devil's staircase for the mode locking intervals which is ordered by the Farey tree.
  - Hausdorff dimension $D_0 = 0.87$ for the unlocked intervals at the critical line which corresponds to $K = 1$ in the circle map.
  - Nontrivial scaling ($\tilde{a} = -1.289$) near the golden mean winding number.

In the following examples, we will show how this program actually works.

Driven Pendulum

One of the simplest physical systems whose dynamical description has been reduced to a circle map (Jensen et al., 1983 a, 1984) is a periodically driven pendulum with an additional constant external torque B which is described by the differential equation:

$$\ddot{\theta} + \gamma \dot{\theta} + \sin \theta = A \cos (\omega t) + B \qquad (6.58)$$

Naive discretization of the time derivative in eq. (6.58) yields for $\theta_n = \theta(t = (2\pi/\omega) \cdot n)$:

$$\theta_{n+1} - 2\theta_n + \theta_n + (1 - b)(\theta_n - \theta_{n-1}) + K \sin \theta_n = (1 - b)\Omega \qquad (6.59)$$

where $(1 - b) = \gamma \dfrac{2\pi}{\omega}$; $\quad \Omega = \dfrac{2\pi}{\omega}(A + B)/\gamma$; $\quad K = \dfrac{2\pi}{\omega}$.

This is, for $r_n = \theta_n - \theta_{n-1} + \Omega$, equivalent to the dissipative circle map:

$$\theta_{n+1} = \theta_n + \Omega - K \sin \theta_n + b r_n \qquad (6.60\,\text{a})$$

$$r_{n+1} = b r_n - K \sin \theta_n . \qquad (6.60\,\text{b})$$

Eqs. (6.59 60) make it plausible that the pendulum has something to do with the circle map, but they do not, of course, establish a rigorous connection. A numerical proof has been given by Jensen et al. (1983 a, 1984) who solved eq. (6.58) by using a computer. Fig. 113 shows that subsequent values of the angles $\theta_n = \theta(t = (2\pi/\omega)n)$; $n = 0, 1, 2 \ldots$ taken at integer multiples of the driving period $2\pi/\omega$ yield, for special parameter values, a one dimensional circle map.

Let us briefly comment on how mode locking shows up in the solutions of eq. (6.58). Mode locking implies:

$$\theta(t_0 + qT) - \theta(t_0) = 2\pi \cdot p \qquad (6.61)$$

where $T = \dfrac{2\pi}{\omega}$. This yields

$$\langle \dot{\theta} \rangle \equiv \frac{1}{qT} \int_{t_0}^{t_0 + qT} dt\, \dot{\theta} = \frac{p}{q} \cdot \omega \qquad (6.62)$$

i.e. the mode locking state of the pendulum is characterized by the fact that its averaged angular velocity $\langle \dot{\theta} \rangle$ is a rational multiple of the external driving frequency ω.

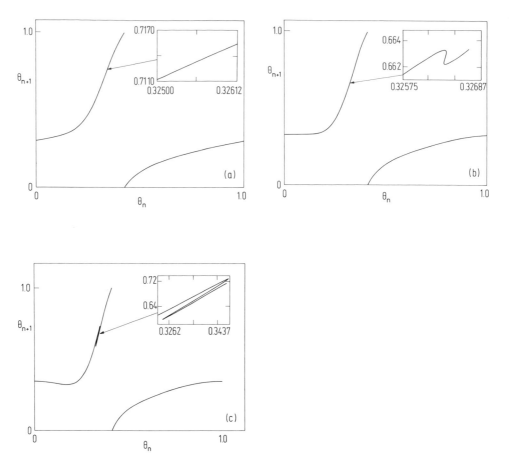

Fig. 113: Poincaré maps obtained by numerical integration of the pendulum equation (6.58), $\theta_n = \theta(t = [2\pi/\omega]n)$. 25 000 consecutive points have been plotted; the first 1000 points have been omitted. Parameters: $A = 1.0$, $\omega = 1.76$. a) $\gamma = 1.576$, $\Omega = 1.4$, the function $f(\theta_n)$ increases monotonically, and the inset is a magnification emphasizing the one dimensional character of the map. b) $\gamma = 1.253$, $\Omega = 1.2$, the map develops a cubic inflection point, indicating the transition to chaos. The inset shows an enlargement around the inflection point. c) $\gamma = 1.081$, $\Omega = 1.094$, the map develops a local minimum and wiggles (insets) indicating chaotic behavior. (After Jensen et al., 1984.)

Furthermore, the universal Hausdorff dimension $D = 0.87$ of the Cantor set derived from the space between mode locked plateaus has also been measured directly in an electronic simulation of a driven pendulum by Yeh et al. (1984). They evaluated the following quantities:

$S(l)$ = total length of all mode locking steps larger than l

$[1 - S(l)]/l = N(l)$ = number of intervals of size l needed to cover the unlocked holes.

From $N(l)$ they obtained, via $\lim_{l \to 0} N(l) \propto l^{-D}$, the Hausdorff dimension D_0 shown in Fig. 114.

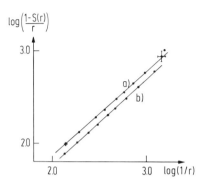

Fig. 114: Plots of log $\{[1 - S(r)]/r\}$ versus log $(1/r)$, the slopes give the fractal dimension D_0. a) $\omega = 2.9$, $A = 1.06$, $1/8 \leq B \leq 1/5$ yields $D_0 = 0.91$ based on 91 steps. b) $\omega = 1.58$, $A = 0.63$, $1/5 \leq B \leq 1/3$ yields $D_0 = 0.92$ based on 45 steps. (After Yeh et al., 1984.)

Let us also call attention to the colored plates, XVI and XVII, at the beginning of this book which show the parameter dependence of the largest Liapunov exponents of a driven pendulum and the corresponding quantity for the circle map. One sees how, in both cases, the Arnold tongues develop and finally merge together as the nonlinearity parameter is increased.

Since eq. (6.58) also describes externally driven Josephson junctions and charge density waves under the influence of an dc and ac electric field (Jensen et al., 1984), one expects that the dynamical behavior of these systems can also be modeled by one dimensional circle maps. This has indeed been partly confirmed experimentally (see References to this section).

Electrical Conductivity in Barium Sodium Niobate

Another fine example, where the circle map and the devil's stair case (associated to mode locking) have been observed experimentally, is the Barium Sodium Niobate crystal that we described already on page 152 (Martin and Martienssen, 1986). The voltage across the crystal displays, under the influence of a constant dc current, spontaneous oscillations that can be modulated by an additional ac current as shown in Fig. 115.

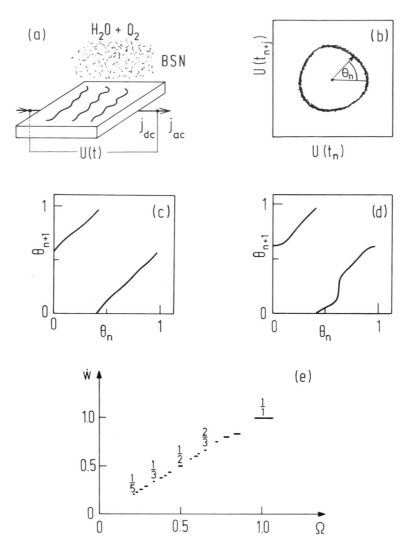

Fig. 115: a) BSN crystal in humidified oxygen atmosphere at a temperature of 535 °C with an ac current density j_{ac} superimposed onto a constant dc current density j_{dc}; also indicated are the "domains" shown in plate V at the beginning of this book. b) Poincaré map constructed from the measured voltage signal. c) and d): the circle map (c) constructed from the measured voltage (a) becomes nonlinear (d) if the dc current density is increased. e) Mode locked states, measured by varying the driving frequency, display a devil's staircase behavior near the transition to chaos. (After Martin and Martienssen, 1986.)

Dynamics of Cardiac Cells

It has been found by M. R. Guevara, L. Glass, and A. Shrier (1981) that circle maps are also relevant for explaining the dynamics of cardiac cells. Fig. 116 shows the temporal behavior of the transmembrane electric potential from an aggregate of embryonic chick heart cells, which beat spontaneously. If the system is *periodically stimulated* via a current pulse through a microelectrode, the nature of the response depends on the interstimulus interval. The main idea is to *reduce this response to a single stimulus* by constructing an appropriate circle map.

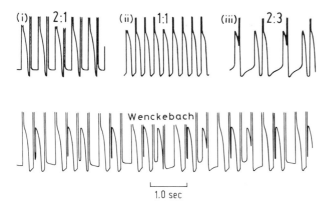

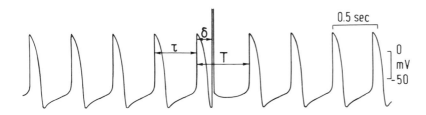

Fig. 116: Influence of periodic stimulation as a function of the interstimulus interval t_s: a) Stable phase locked pattern (i) 2:1 t_s = 210 msec; (ii) 1:1, t_s = 240 msec; (iii) 2:3 t_s = 600 msec. b) Irregular dynamics displaying the Wenckebach phenomenon, t_s = 280 msec. (After Guevara et al., 1981; copyright 1981 by the AAAS.)

Fig. 117: Time course of the transmembrane electrical potential from an aggregate of embryonic heart cells. Left: Spontaneous pulses. Right: After administration of a brief depolarizing stimulus (off-scale response) which occurs δ msec after the action potential upstroke. The graph sharply rises, and the spontaneous-state period τ is shifted to a new value T. (From Guevara et al., 1981; copyright 1981 by the AAAS.)

Fig. 117 shows that the influence of a single pulse changes the period of the spontaneous beats from τ to T. The assumption is now that their ratio T/τ depends only on the phaseshift $\theta = \delta/\tau$ of the stimulus with respect to the natural signal, that is,

$$T/\tau = g(\theta). \tag{6.63}$$

This assumption is supported by the experimentally determined function $g(\theta)$ displayed in Fig. 118.

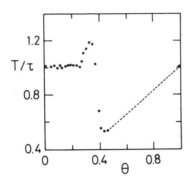

Fig. 118: The function $g(\theta)$ defined in eq. (6.63), as experimentally determined for embryonic chick heart cell aggregates (from Guevara et al., 1981; copyright 1981 by the AAAS).

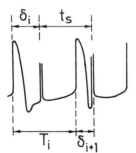

Fig. 119: Graphical demonstration of the relation
$T_i + \delta_{i+1} = \delta_i + t_s$ for $T_i < \delta_i + t_s < T_i + \tau$.

Next we consider a train of stimuli separated by a uniform time interval t_s. Consultation of Fig. 119 leads to the relation

$$\delta_{i+1} + T_i = \delta_i + t_s. \tag{6.64}$$

Division by τ, and assuming that the influence of a single stimulus decays sufficiently fast such that eq. (6.63) holds for every i, yields the phase relationship:

$$\theta_{i+1} = \theta_i + \Omega - g(\theta_i); \quad \Omega \equiv t_s/\tau \tag{6.65}$$

which has the form of a circle map (see Fig. 120) where the rate of rotation $\Omega = t_s/\tau$ is set by the interstimulus distance t_s.

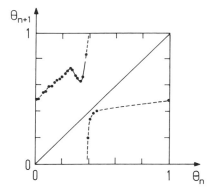

Fig. 120: Experimentally determined circle map that describes the dynamics of beating chicken heart cell aggregates. This graph is obtained by using $g(\theta)$ from Fig. 118 in eq. (6.65). (From Guevara et al., 1981; copyright 1981 by the AAAS.)

Using $g(\theta)$ from Fig. 118, eq. (6.65) has been used to successfully predict the response to a train of stimuli as a function of t_s (see Fig. 121). The so-called Wenckebach phenomenon in Fig. 116c (i.e., the gradual prolongation of the time between a stimulus and the subsequent action potential until an active potential is skipped either irregularly or in a phase locked pattern) occurs also in human electrocardiograms (Fig. 122). There the external stimulus is replaced by the stimulus provided by the sinoatrial node. It appears, therefore, from the results in Fig. 121 that circle maps provide a promising tool for the investigation of human cardiac dysrhythmia.

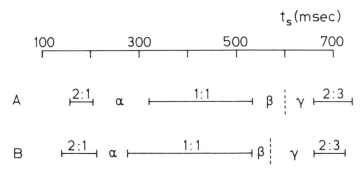

Fig. 121: Experimentally determined and theoretically computed responses to periodic stimulation of period t_s with the same pulse durations and amplitudes as in Fig. 116a). a) Experimentally determined dynamics: 2:1, 1:1, 2:3 mode locking regions and three zones α, β, γ of complicated dynamics. b) Theoretically predicted dynamics obtained via eq. (6.65). (After Guevara et al., 1981; copyright 1981 AAAS.)

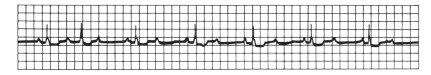

Fig. 122: Human electrocardiogram showing one 3:2 Wenckebach cycle followed by five 2:1 cycles (after Glass et al., 1981).

Forced Rayleigh-Bénard Experiment

Another example where the global metric properties of the attractor which occurs at the transition from quasiperiodicity to chaos at the golden mean winding number, have been measured in some detail is a forced Rayleigh Bénard experiment by Jensen et al. (1985). One uses mercury as a fluid in a small Rayleigh-Bénard cell that supports two convection rolls. The Rayleigh number is chosen in a range where the convection is oscillatory in time. A second frequency is introduced by passing an ac current through the fluid whose amplitude and frequency serve as control parameters. Fig. 123 a shows the reconstructed experimental orbit obtained at the point the breakdown of the torus which has a golden mean winding number. The dots in Fig. 123 b are the experimental points derived from the data shown in Fig. 123 a, and the full line is the $f(\alpha)$ curve obtained from the time series of the circle map at $K = 1$, $w^* = (\sqrt{5} - 1)/2$.

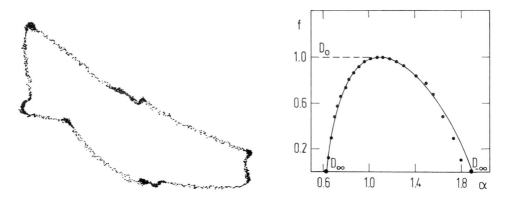

Fig. 123: a) The experiment attractor of a forced Rayleigh-Bénard system in two dimensions. 2500 points are plotted. Note the variation in the density of points on the attractor. Part of this variation is, however, due to the projection of the attractor onto the plane. The attractor is nonintersecting in three dimensions, in which it was embedded for the numerical analysis. In the absence of experimental noise, the points should fall on a single curve. The smearing of the observed data set is mostly due to the slow drift in the experimental system during the run over about 2 hours. b) full line: $f(\alpha)$ curve obtained from the iterates of the circle map eq. (6.13) at $K = 1$ and the golden mean winding number; dots: $f(\alpha)$ values obtained from the experimental data in a). (After Jensen et al., 1985.)

The agreement between both sets of data is rather obvious and leads to the conclusion that the experimental data in Fig. 123 a, which look not at all like a smooth circle, and the iterates of the circle map (6.13) belong, from the metric point of view, to the same universality class.

We note that this experiment yields also via $D_{-\infty}$ the first measurement of the nontrivial scaling parameter $\tilde{a}$ of the circle map. $D_{-\infty}$ has, at the transition from quasiperiodicity to chaos, the value

$$D_{-\infty} = -\frac{\log w^*}{\log \tilde{a}} = 1.8980\ldots \tag{6.67}$$

which is obtained for circle maps, in analogy to eq. (5.81), by replacing the ratio in the number of subsequent cycles (which is 2 for period doubling) by $F_{n+1}/F_n \simeq 1/w^*$ and using $\tilde{a}$ instead of a.

6.4   Routes to Chaos

Table 12 summarizes the three different routes to chaos which we have discussed up to now.

Table 12: Summary of three main routes to chaos.

Feigenbaum	Manneville-Pomeau	Ruelle-Takens-Newhouse
Pitchfork bifurcation	Tangent bifurcation	Hopf bifurcation
Bifurcation diagrams		
Main phenomena		
Infinite cascade of period doublings with universal scaling parameters	Intermitted transition to chaos. The laminar phase has a duration $(r - r_c)^{-1/2}$	After three bifurcation, strange attractor "probable".
Experiments		
Bénard experiment Taylor experiment Driven nonlin. oscill. Chemical reactions Optical instabilities	Bénard experiment Josephson junction Chemical reactions Lasers	Bénard experiment Taylor experiment Nonlin. conductors

But this table should only be considered as a first approximation to the true variety of transition scenarios. (Let us only recall that we have already discussed three types of intermittency). While it is natural to focus on common features, it would be premature to make sweeping generalizations about routes to chaos, and it should be emphasized that the range of dynamical behavior observed is quite large.

This situation arises, on the one hand, because experiments on hydrodynamic systems (Bénard and Taylor instability) depend sensitively on the *aspect ratios* (i. e. the ratio of the cell dimensions in the Bénard experiment, and the ratio of the width between the inner and outer cylinder and the height of the cylinder in the Taylor experiment) such that, for a given set of control parameters, *one can have more than one stable state*. On the other hand, new types of transitions are possible when one has *more than one control parameter* (Swinney, 1983). Let us finally present a transition to chaos not mentioned above.

Crises

Crises are collisions between a chaotic attractor and a coexisting unstable fixed point or periodic orbit. Grebogi, Ott and York (1983 b) were the first to observe that such collisions lead to *sudden* changes in the chaotic attractor. A simple example occurs in the period-three window of the logistic map in Fig. 51, where three stable and three unstable fixed points are generated by tangent bifurcations. Fig. 124 shows that the unstable fixed points, having entered the chaotic regions, immediately repel the trajec-

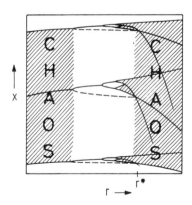

Fig. 124: Detail of the bifurcation diagram in the region of the period-three tangent bifurcation. The dashed curves denote the unstable period-three orbit created at the tangent bifurcation; the crises occur at r^* (schematic, after Grebogi et al., 1983 b).

tory out of the sub-band in such a way that the regions between the bands are also filled chaotically. Similar crises also occur in two- and three-dimensional maps and in three-dimensional flows.

As the discontinuity is approached, one often finds transient chaos, i.e. seemingly chaotic orbits which decay exponentially towards periodic orbits with a decay rate that follows a power law of the distance (in parameter space) from the discontinuity. It has been conjectured by Grebogi et al. (1983 b) that "almost all" sudden changes in chaotic attractors are due to crises.

7 Regular and Irregular Motion in Conservative Systems

Up to now we have exclusively studied dissipative systems for which volume elements in phase space shrink with increasing time. Although there are many physical realizations of dissipative systems, which range from the onset of turbulence in fluids to electronic circuits, there exists another large class of physical systems for which chaotic motion has been found (by Poincaré, 1982) before the discovery of the strange attractor for dissipative systems (Lorenz, 1963): These are the conservative systems which encompass all dynamical systems of classical mechanics.

Because there already exist excellent review articles by Berry (1978) and Helleman (1980) and a recent book by Lichtenberg and Liebermann (1982) on this subject, our presentation will be rather brief (as compared to six chapters on dissipative systems).

In the following, conservative systems are considered to be either systems that follow Hamilton's equations of motion,

$$\dot{\vec{q}} = \frac{\partial H}{\partial \vec{p}}, \quad \dot{\vec{p}} = -\frac{\partial H}{\partial \vec{q}} \tag{7.1}$$

and for which, volume elements in phase space are conserved because of Liouville's theorem,

$$\operatorname{div} \vec{j} = \operatorname{div}(\dot{\vec{q}}, \dot{\vec{p}}) = \sum_i \left(\frac{\partial^2 H}{\partial q_i p_i} - \frac{\partial^2 H}{\partial p_i q_i} \right) = 0 \tag{7.2}$$

or, in a more general sense, volume preserving, discrete maps.

The fact that volumes do not change in conservative systems implies immediately that they display (in contrast to dissipative systems) no attracting regions in phase space, i.e. no attracting fixed points, no attracting limit cycles, and no strange attractors (see Fig. 125 and Appendix G). Nevertheless, in conservative systems one also finds chaos with a positive K-entropy, i.e. there are "strange" or "chaotic" regions in phase space, but they are not attractive and can be densely interweaved with regular regions.

a)   b)

Fig. 125: a) In dissipative systems trajectories are attraced to fixed point, and volume shrinks, b) In conservative systems the points rotate around an elliptic fixed point, volume is conserved.

We now present some motivation for the study of conservative systems and then give an overview over the rest of this chapter.

For some time, attention has shifted from the calculation of individual orbits to consideration of the qualitative properties of families of orbits, as shown in Fig. 126. Today, we are mainly interested in the long-time behavior of conservative systems. There are several reasons for this:

a) We should, for example, be able to answer the question whether the solar systems and the galaxy are stable under mutual perturbations of their constituents, or whether they will eventually collapse or disperse to infinity. The long-time limit involved here is of the order of the age of the universe. But "long" times are much shorter in the storage rings used for high energy physics or in fusion experiments, where the particles make many revolutions in fractions of a second. In such systems irregular or chaotic motion is to be avoided at all costs, and this is only possible if the long-time behavior of these (conservative) systems is known.

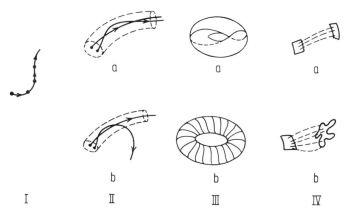

I   II   III   IV

Fig. 126: Problems of increasing globality in classical mechanics. I. Step by step integration of the equations of motion. II. a) Local stability; b) local instability. III. Topological nature of complete trajectories: a) periodic motion on a torus; b) motion on a torus with irrational frequency ratios. IV. Types of flow in phase space: a) non mixing; b) mixing. (After Balescu, 1975.)

b) another point concerns the foundations of statistical mechanics, where no attempt is made to follow the detailed motion of all constituents of a complicated manybody problem. Instead, the ergodic hypothesis is made, i. e. one assumes that in the course of time the system explores the entire region of phase space allowed (the energy surface) and eventually covers this region uniformly. Time averages can then be replaced by simpler phase-space averages. But is the ergodic hypothesis correct? To answer this question, the long-time behavior of Hamiltonian systems with N degrees of freedom in the limit $N \to \infty$ (and N/volume = constant) must be known.

In the first part of this section, we consider the classical mechanis of simple Hamiltonian systems with a few degress of freedom and show that in most cases their motion in phase space is extremely complicated and neither regular nor simply ergodic. In other words, it will be shown that the regular motion treated in most textbooks on classical mechanics is an exception and rather uncommon.

In the second part, we discuss some simple model systems which behave ergodically although they have only a few degrees of freedom. Finally, a classification scheme for chaotic behavior in conservative systems is described.

7.1 Coexistence of Regular and Irregular Motion

In the following, we investigate the stability of the trajectories of a nonintegrable Hamiltonian system in the long-time limit. For this purpose, we start from an integrable Hamiltonian and consider the effect of a small nonintegrable perturbation.

Integrable Systems

A Hamiltonian $H_0'(\vec{p}, \vec{q})$ is called integrable if one can find a canonical transformation $S(\vec{q}, \vec{J})$ to new variables $\vec{\theta}, \vec{J}$:

$$\vec{q}, \vec{p} = \frac{\partial S(\vec{q}, \vec{J})}{\partial \vec{q}} \leftrightarrow \vec{J}, \vec{\theta} = \frac{\partial S(\vec{q}, \vec{J})}{\partial \vec{J}} \tag{7.3}$$

such that in the new coordinates the Hamiltonian depends only on the new momenta $\vec{J}$, i. e., $S(\vec{q}, \vec{J})$ is a solution of the *Hamilton-Jacobi equation* (see, e. g., Arnold, 1978):

$$H_0'\left[\vec{q}, \frac{\partial S(\vec{q}, \vec{J})}{\partial \vec{q}}\right] = H_0(\vec{J}) \tag{7.4}$$

and the equations of motion in the action-angle variables $\vec{J}$ and $\vec{\theta}$

$$\dot{\vec{J}} = -\frac{\partial H_0}{\partial \vec{\theta}} = 0 \tag{7.5}$$

$$\dot{\vec{\theta}} = \frac{\partial H_0}{\partial \vec{J}} = \vec{\omega}(\vec{J})$$

can easily be integrated to

$$\vec{J} = \text{const.}$$
$$\vec{\theta} = \vec{\omega} \cdot t + \vec{\delta} . \tag{7.7}$$

One of the simplest examples for an integrable system is a harmonic oscillator that has the Hamiltonian

$$H_0' = \frac{1}{2}(p^2 + \omega^2 q^2) . \tag{7.8}$$

The Hamilton-Jacobi equation (7.4) then becomes

$$\frac{1}{2}\left[\left(\frac{\partial S}{\partial q}\right)^2 + \omega^2 q^2\right] = H_0(J) \tag{7.9}$$

$$\rightarrow \frac{\partial S}{\partial q} = \sqrt{2H_0 - \omega^2 q^2} \tag{7.10}$$

and J is determined by

$$J = \frac{1}{2\pi}\oint \frac{\partial S}{\partial q}\,dq = \frac{H_0(J)}{\omega} \tag{7.11}$$

$$\rightarrow H_0(J) = J\omega \tag{7.12}$$

where the integral has been taken over one cycle of q.
  The equations of motion in the action-angle variables are

$$\dot{J} = \frac{\partial H_0}{\partial \theta} = 0 \rightarrow J = \text{const} \tag{7.13a}$$

$$\dot{\theta} = \frac{\partial H_0}{\partial J} = \omega \rightarrow \theta = \omega t + \delta . \tag{7.13b}$$

The motion in the variables p and q is obtained from

$$\theta = \frac{\partial S}{\partial J} = \frac{\partial}{\partial J} \int dq \sqrt{2H_0 - \omega^2 q^2} = \arccos\left(q\sqrt{\frac{\omega}{2J}}\right) \tag{7.14}$$

$$\rightarrow q = \sqrt{\frac{2J}{\omega}}\cos\theta \tag{7.15}$$

and

$$p = \frac{\partial S}{\partial q} = -\sqrt{2J\omega}\sin\theta . \tag{7.16}$$

The corresponding trajectory in phase space is an ellipse that becomes a circle with polar coordinates $\sqrt{J}$ and θ after proper rescaling. Comparing eqns. (7.7) and (7.13) one sees that the equations of motion (in action-angle variables) of any integrable system with n degrees of freedom are practically the same as those of a set of n uncoupled harmonic oscillators. The only difference is that in a general integrable system the frequencies ω_i are still functions of the actions J_i whereas they are independent of J_i for harmonic oscillators. The existence of n integrals of the motion $(J_1 \ldots J_n)$ confines the trajectory in the 2n-dimensional phase space $(q_1 \ldots q_n, p_1 \ldots p_n)$ of an integrable system to an n-dimensional manifold which has − in analogy to a circle for a harmonic oscillator with $n = 1$ and a torus for two harmonic oscillators with $n = 2$ − the topology of an n-torus.

In the following, we will confine ourselves to $n = 2$, but most results can be extended to more degrees of freedom. Fig. 127 shows the motion of an integrable system with two degrees of freedom (i. e. with a 4-dimensional phase space) on a torus. Closed orbits occur only if

$$n\Delta\theta_2 = 2\pi \cdot m, \text{ i. e. } \frac{\omega_2}{\omega_1} = \frac{m}{n} = \text{rational} ; m, n = 1, 2, 3 \ldots . \tag{7.17}$$

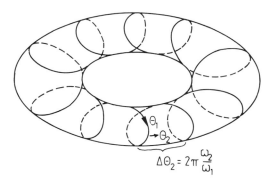

Fig. 127: Torus in phase space.

For irrational frequency ratios, the orbit never repeats itself but approaches every point on the two-dimensional manifold infinitesimally close in the course of time. In other words, the motion is ergodic on the torus. (Note that the dimension 2 of the torus is different from the dimension 3 of the manifold defined by $H(\vec{p}, \vec{q}) = E = \text{const.}$)

Perturbation Theory and Vanishing Denominators

Let us now add to H_0 a perturbation εH_1 and see how it effects the previously regular motion; that is, we consider the Hamiltonian

$$H(\vec{J}, \vec{\theta}) = H_0(\vec{J}) + \varepsilon H_1(\vec{J}, \vec{\theta}) \tag{7.18}$$

(where we expressed H_1 in the action-angle variables $\vec{J} = (J_1, J_2)$, $\vec{\theta} = (\theta_1, \theta_2)$ of the unperturbed system), and we try to solve the Hamilton-Jacobi equation

$$H\left[\frac{\partial S}{\partial \vec{\theta}}, \vec{\theta}\right] = H_{00}(\vec{J}') . \tag{7.19}$$

Writing the generating function S as

$$S(\vec{J}', \vec{\theta}) = \vec{\theta} \cdot \vec{J}' + \varepsilon S_1(\vec{J}', \vec{\theta}) \tag{7.20}$$

and expanding H to order ε, we obtain

$$H_0(\vec{J}') + \varepsilon \frac{\partial H_0}{\partial \vec{J}} \cdot \frac{\partial S_1(\vec{J}', \vec{\theta})}{\partial \vec{\theta}} + \varepsilon H_1(\vec{J}', \vec{\theta}) + O(\varepsilon^2) = H_{00}(\vec{J}') \tag{7.21}$$

S_1 is determined by requiring that the left-hand side in (7.21) is independent of $\vec{\theta}$, i. e.

$$\vec{\omega} \cdot \frac{\partial S_1(\vec{J}', \vec{\theta})}{\partial \vec{\theta}} = -H_1(\vec{J}', \vec{\theta}) \tag{7.22}$$

where $\vec{\omega} = \partial H_0/\partial \vec{J}'$ are the frequencies of the unperturbed system. Eq. (7.21) can be solved by expanding S_1 and H_1 (both being periodic in the components of $\vec{\theta}$) into Fourier series:

$$S_1(\vec{J}', \vec{\theta}) = \sum_{\vec{K} \neq 0} S_{1,\vec{K}}(\vec{J}') e^{i\vec{K} \cdot \vec{\theta}} \tag{7.23 a}$$

$$H_1(\vec{J}', \vec{\theta}) = \sum_{\vec{K} \neq 0} H_{1,K}(\vec{J}') e^{i\vec{K} \cdot \vec{\theta}} \tag{7.23 b}$$

with $\vec{K} = 2\pi(n_1, n_2)$; n_1, n_2 integers.

Using both expressions in (7.22) and comparing equal Fourier components finally yields

$$S(\vec{J}', \vec{\theta}) = \vec{\theta} \cdot \vec{J}' + i\varepsilon \sum_{\vec{K} \neq 0} \frac{H_{1,K}(\vec{J}')}{\vec{K} \cdot \vec{\omega}(\vec{J}')} e^{i\vec{K} \cdot \vec{\theta}} . \tag{7.24}$$

Equation (7.24) shows that S diverges for

$$\omega_1 n_1 + \omega_2 n_2 = 0 , \text{ i.e. } \frac{\omega_1}{\omega_2} = -\frac{n_2}{n_1} = \text{rational} . \tag{7.25}$$

This is the famous problem of vanishing denominators. It shows that the system cannot be integrated by perturbation theory for rational frequency ratios because of strong resonances, and it seems that it can at most be integrated for irrational values of ω_1/ω_2 if the perturbation series in ε converges.

In the following we consider two problems:

− What happens if an integrable system with ω_1/ω_2 close to an *irrational* value is perturbed by εH_1?

− What happens under a perturbation εH_1 to the tori of a system for which ω_1/ω_2 has a *rational* value?

Stable Tori and KAM Theorem

The first question is answered by a celebrated theorem of Kolmogorov (1954), Arnold (1963), and Moser (1967), the so-called KAM theorem which we quote here for $n = 2$, without proof. (The theorem holds for an arbitrary number n of degrees of freedom and proofs can be found in the quoted references.) The theorem states that if, among other technical conditions, the Jacobian of the frequencies is nonzero, i.e.

$$\left| \frac{\partial \omega_i}{\partial J_j} \right| \neq 0 \tag{7.26}$$

then those tori, whose frequency ratio ω_2/ω_1 is sufficiently irrational such that

$$\left| \frac{\omega_1}{\omega_2} - \frac{m}{s} \right| > \frac{k(\varepsilon)}{s^{2.5}} \qquad (k(\varepsilon \to 0) \to 0) \tag{7.27}$$

holds (m and s are mutually prime integers), are stable under the perturbation εH_1 in the limit $\varepsilon \ll 1$.

It is important to note that the set of frequency ratios, for which (7.27) holds and for which the motion is therefore regular, even after the perturbation, has a nonzero measure. This follows because the total length L of all intervals in $0 \leq \omega_1/\omega_2 \leq 1$, say, for which (7.27) *does not hold* can be estimated as

$$L < \sum_{s=1}^{\infty} \frac{k(\varepsilon)}{s^{2.5}} \cdot s = k(\varepsilon) \sum_{s=1}^{\infty} s^{-1.5} = \text{const.} \cdot k(\varepsilon) \to 0 \text{ for } \varepsilon \to 0 . \quad (7.28)$$

Here $k(\varepsilon)/s^{2.5}$ is the length of an interval around the rational m/s where (7.27) does not apply, and s is the number of m values with $m/s \leq 1$ (see Fig. 128).

Fig. 128: Intervals of lengths $k(\varepsilon)/s^{2.5}$ contributing to L.

Eq. (7.28) means that the set of frequency ratios, for which (under a perturbation by εH_1) the original motion on the torus is only slightly disturbed into the motion of a deformed torus, has the finite measure $1 - \text{const.} \cdot k(\varepsilon)$. But, on the ω_1/ω_2 axis, this set has holes around every rational ω_1/ω_2.

For large enough ε the perturbation εH_1 destroys *all* tori. The last KAM torus which will be destroyed is the one for which the frequency ratio is the "worst irrational number" $\omega_1/\omega_2 = (\sqrt{5} - 1)/2$ (see Sect. 6.2). The destruction of this KAM torus shows some similarity to the Ruelle-Takens route to chaos in dissipative systems. It has indeed been found by Shenker and Kadanoff (1982) and McKay (1983) who studied the conservative version ($b = 1$) of the map (6.12) of the annulus onto itself that the decay of the last KAM trajectory shows scaling behavior and universal features.

Unstable Tori and Poincaré-Birkhoff Theorem

Let us now discuss the situation when ω_1/ω_2 is rational. We will show that in this case the original torus decomposes into smaller and smaller tori. Some of these newly created tori are again stable according to the KAM theorem. But, between the stable tori, the motion is completely irregular.

It is convenient to visualize what happens (to H_0 under a perturbation εH_1) in a Poincaré map that is, in general, defined by the intersection points of the orbit with a hyperplane in phase space. For the case in hand, we consider the intersections with the q_1, p_1 plane S shown in Fig. 129, which define an area-preserving two-dimensional map

$$r_{i+1} = r_i; \qquad r_i = r\left(t = i \cdot \frac{2\pi}{\omega_2}\right) \qquad (7.29)$$

$$\theta_{i+1} = \theta_i + 2\pi \frac{\omega_1}{\omega_2}$$

since the point in phase space hits S after a period $2\pi/\omega_2$ during which θ changes by $2\pi\omega_1/\omega_2$.

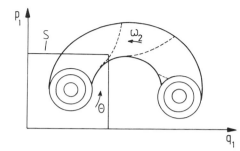

Fig. 129: Poincaré map of orbits on the torus in the plane (q_1, p_1).

The frequency ratio ω_1/ω_2 depends only on the radius r because

$$\left.\begin{aligned} \frac{\omega_1}{\omega_2} &= \frac{\dfrac{\partial H_0(J_1, J_2)}{\partial J_1}}{\dfrac{\partial H_0(J_1, J_2)}{\partial J_2}} = f(J_1, J_2) \\[2em] H_0(J_1, J_2) &= E \rightarrow J_2 = J_2(J_1) \\[1em] J_1 &= \frac{1}{2\pi} \oint p_1 \, dq_1 = \frac{r^2}{2} \end{aligned}\right\} \qquad \frac{\omega_1}{\omega_2} = a(r) \qquad (7.30)$$

(7.30) can therefore be written as

$$\left.\begin{aligned} r' &= r \\ \theta' &= \theta + 2\pi a(r) \end{aligned}\right\} \equiv T\begin{pmatrix} r \\ \theta \end{pmatrix}. \qquad (7.31)$$

This is Moser's twist map (Moser, 1973).

We note that for a rational frequency ratio $r/s = a(r_0)$ every point on the circle r_0, θ_0 is a fixed point of T^s since

$$T^s \begin{pmatrix} r_0 \\ \theta_0 \end{pmatrix} = \begin{cases} r_0 \\ \theta_0 + 2\pi\, \dfrac{r}{s} \cdot s = \theta_0 + 2\pi r. \end{cases} \qquad (7.32)$$

If we now perturb H_0 by εH_1, the twist map becomes

$$\left.\begin{aligned} r_{i+1} &= r_i + \varepsilon f(r_i, \theta_i) \\ \theta_{i+1} &= \theta_i + 2\pi a(r_i) + \varepsilon g(r_i, \theta_i) \end{aligned}\right\} \equiv T_\varepsilon \begin{pmatrix} r_i \\ \theta_i \end{pmatrix} \qquad (7.33)$$

where f and g depend on H_1. As a consequence of Liouville's theorem (which also holds for the Hamiltonian $H_0 + \varepsilon H_1$), the map T_ε is area-preserving.

What can we say now about the fixed points of T_ε? Consider two circles C_+ and C_- between which lies the circle C on which $a = r/s$. On C_+, $a > r/s$ and on C_-, $a < r/s$. T^s therefore maps C_+ anti-clockwise, C_- clockwise, and C not at all (see Fig. 130).

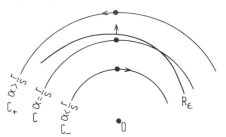

Fig. 130: Action of T^s and T_ε^s on C_+ and C_-.

Under the perturbed map T_ε^s these relative twists are preserved if ε is small enough. Thus, on any radius from 0 there must be one point whose angular coordinate is unchanged by T_ε^s. These radially mapped points make up a curve R_ε close to C.

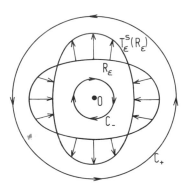

Fig. 131: The curve of radially mapped points R_ε and its image $T_\varepsilon^s (R_\varepsilon)$.

Fig. 131 shows the curve R_ε formed by these points, and its image $T_\varepsilon^s (R_\varepsilon)$ which cuts R_ε in an even number of points because the area enclosed by R_ε and $T_\varepsilon^s (R_\varepsilon)$ must be the same.

The points common to R_ε and $T_\varepsilon^s (R_\varepsilon)$ are the fixed points of T_ε^s, and we can see in Fig. 132 that an alternating sequence of elliptic and hyperbolic fixed points emerges.

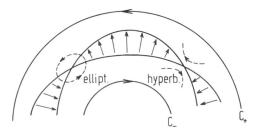

Fig. 132: Alternating hyperbolic and elliptic fixed points of T_ε^s.

This means that the original torus with rational frequency ratio is not completely destroyed under a perturbation, but there remains an even number of fixed points. This is the „Poincaré-Birkhoff theorem" (Birkhoff, 1935).

Let us first consider the elliptic fixed points which are surrounded by rotating points (see Figs. 125, 132). The corresponding orbits are the Poincaré sections of smaller tori for which all our arguments can be repeated; that is, some of these smaller

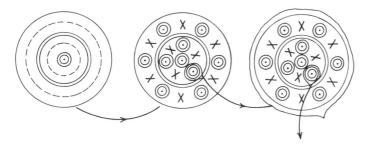

Fig. 133: Tori with rational frequency ratio decay into smaller and smaller tori, and the pattern of newly created elliptic and hyperbolic fixed points shows self-similarity.

tori are again stable according to the KAM theorem and other tori decompose into smaller ones according to the Poincaré-Birkhoff theorem. This gives rise to the self-similar structure in Fig. 133.

Homoclinic Points and Chaos

Which role do the hyperbolic fixed points play? Fig. 134 shows that, near a hyperbolic fixed point H, the motion becomes unstable, and orbits are driven away from it, in contrast to the stable rotational motion around an elliptic fixed point.

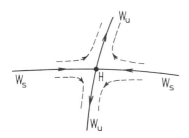

Fig. 134: Hyperbolic fixed point H with stable (W_s) and unstable (W_u) lines.

The stable (W_s) and unstable (W_u) lines which lead to or emanate from H behave highly irregularly since:

a) They cannot intersect themselves (otherwise the motion on a trajectory in phase space would not be unique for a given set of initial conditions),

b) but W_u can intersect W_s at a so-called homoclinic point (see Fig. 135).

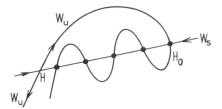

Fig. 135: Homoclinic points H_0 are the intersections of W_u and W_s.

Because the map T_ε^s is continuous, and a homoclinic point is no fixed point, repeated application of T_ε^s produces new homoclinic points. Furthermore, T_ε^s must be applied an infinite number of times to approach the hyperbolic fixed point H along W_s (Appendix G.) Between each homoclinic point H_0 and H there is, therefore, an infinite number of other homoclinic points; that is, the curves W_u and W_s form an extremely complex network.

Summarizing: If we disturb the regular orbits of an integrable system on a torus in phase space by adding a nonintegrable perturbation, then, depending on the different initial conditions (different $\vec{J}, \vec{\delta}$ in (6.7)) imply different ω_1/ω_2 since $\vec{\omega} = \vec{\omega}(\vec{J})$, regular or completely irregular motion results. Although the measure of initial conditions, which lead to regular motion, is nonzero due to the KAM theorem, for every rational frequency ratio (which are densely distributed along the real axis) one obtains smaller and smaller stable tori and irregular orbits due to the hyperbolic fixed points. Thus, an arbitrarily small change in the initial conditions leads to a completely different long-time behavior; and for the motion in phase space, one obtains the complicated pattern in Fig. 136. It shows that in conservative systems regular and irregular motion are densely interweaved.

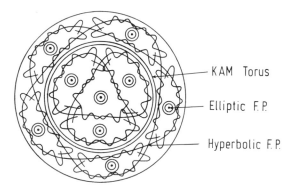

KAM Torus

Elliptic F.P.

Hyperbolic F.P.

Fig. 136: Regular and irregular motion in the phase space of a nonintegrable system.

Finally, we also mention that for area-preserving maps one finds "period doubling", i. e. a successive creation of new pairs of *elliptic* fixed points (Greene et al., 1981). We shall discuss this scenario in Appendix G and show that the corresponding Feigenbaum constants are larger than in the dissipative case.

Arnold Diffusion

So far in this section we have only dealt with systems having two degress of freedom for which the two-dimensional tori stratify the three-dimensional energy surface S_E. The irregular orbits which traverse regions where rational tori have been destroyed are therefore trapped between irrational tori. They can only explore a region of the energy surface which, while three-dimensional, is nevertheless restricted and, in particular, disconnected from other irregular regions, as shown in Fig. 137.

For more degrees of freedom, however, the tori do not stratify S_E (e. g. for three degrees of freedom the tori are three-dimensional, and the energy surface is five-dimensional). The gaps then form one single connected region. This offers the possibility of so-called *"Arnold diffusion"* of irregular trajectories (Arnold, 1964). The existence of invariant tori for perturbed motion is, therefore, not a guarantee of stability of motion for systems with more than two degrees of freedom because irregular wandering orbits that are *not* trapped exist arbitrarily close to the tori.

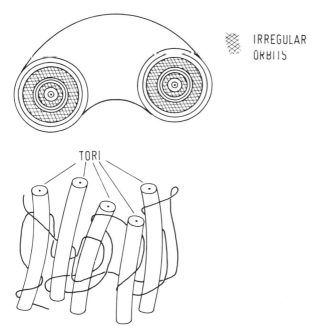

IRREGULAR
ORBITS

Fig. 137: Trapping of irregular orbits between stable KAM tori for a system with two degrees of freedom.

TORI

Fig. 138: Arnold diffusion for Hamiltonian systems with more than two degrees of freedom (schematically).

Examples of Classical Chaos

Finally, we present some experimental evidence for the coexistence of regular and irregular motion. Fig. 139 shows the Poincaré map in S for the nonintegrable Hénon-Heiles system,

$$H = \frac{1}{2} \, p_1{}^2 + q_1^2 + p_2^2 + q_2^2 + \left[q_1^2 q_2 - \frac{q_2^3}{3} \right] \tag{7.34}$$

which consists of an integrable pair of harmonic oscillators coupled by nonintegrable cubic terms (Hénon, Heiles, 1964). The left-hand column shows the surfaces of section generated by eighth-order perturbation theory for various energies (after Gustavson, 1966). The right-hand side are the computed intersections of the trajectory with S. For $E = 1/24$ and E 1/12, the mapping plane is covered with the intersections of (somewhat deformed) tori which signal regular motion and which are identical with those given by perturbation theory. Above $E = 1/9$, however, most, but not all, tori are destroyed, and all the dots which appear to be random are generated by one trajec-

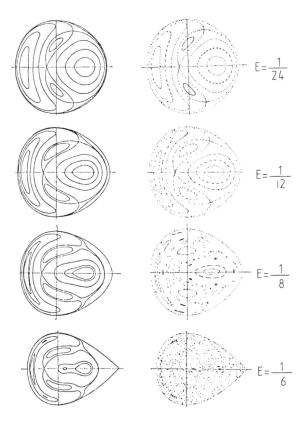

Fig. 139: Poincaré maps for the Hénon-Heiles system (after Berry, 1978).

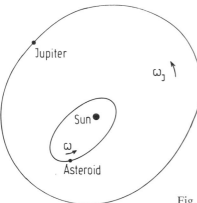

Fig. 140: Perturbation of an asteroid's motion by Jupiter.

tory as it crosses S. The figure for $E = 1/8$ clearly shows the coexistence of regular and irregular motion.

As a further example, we consider the motion of an asteroid around the sun, perturbed by the motion of Jupiter, as shown in Fig. 140.

This three-body problem is nonintegrable, and according to eqns. (7.24–25) we expect that the asteroid motion becomes unstable if the ratio of the unperturbed frequency of the asteroid motion ω and the angular frequency of Jupiter ω_J becomes rational. Fig. 141 illustrates that, in fact, gaps occur in the asteroid distribution for rational ω/ω_j. On the other hand, the existence of stable asteroid orbits ($f \neq 0$) can be considered as a confirmation of the KAM theorem.

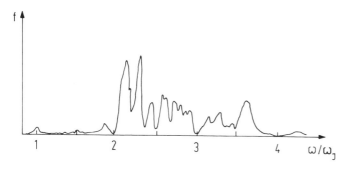

Fig. 141: Fraction f of asteroids in the belt between Mars and Jupiter as a function of ω/ω_j (after Berry, 1978).

A second sort of solar-system gaps occurs in the *rings of Saturn*. In this system Saturn is the attractor; the perturber is any of the inner satellites, and the rest masses are the ring particles. One major resonance occurs within the "Cassini division" shown on Plate VII at the beginning of the book.

7.2 Strongly Irregular Motion and Ergodicity

In the previous section, we linked the origin of irregular motion in Hamiltonian systems to hyperbolic fixed points in the associated area-preserving maps. If we, therefore, want to construct models for strongly irregular motion, it is natural to search for maps for which all fixed points are hyperbolic.

Cat Map

One example of such a system is Arnold's cat map on a torus which is defined by

$$\left.\begin{array}{l} x_{n+1} = x_n + y_n \bmod 1 \\[2mm] y_{n+1} = x_n + 2y_n \bmod 1 \end{array}\right\} \equiv T\begin{pmatrix} x_n \\ y_n \end{pmatrix} \tag{7.35}$$

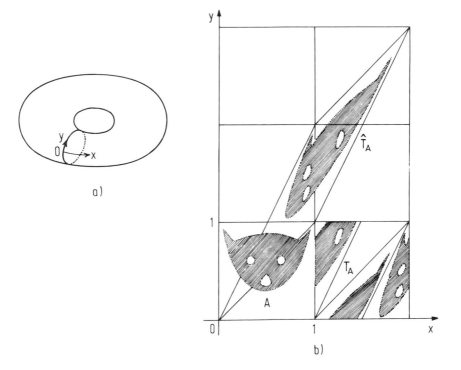

a)

b)

Fig. 142: Action of the map T on a cat on a torus. The torus a) is transposed into the unit square of b). $\hat{T}$ is the map T without restriction to the torus. (After Arnold and Avez, 1968.)

This map is area-preserving because the Jacobian of T is unity, and it has the eigenvalues

$$\lambda_1 = (3 + \sqrt{5})/2 > 1 \quad \text{and} \quad \lambda_2 = \lambda_1^{-1} < 1 \tag{7.36}$$

so that all fixed points of T^n ($n = 1, 2, 3 \ldots$) are hyperbolic. Any point on the torus for which x_0 and y_0 are rational fractions is a fixed point of T^n for some n (e. g. $(0, 0)$ is a fixed point of T, and $(2/5, 1/5)$ and $(3/5, 4/5)$ are fixed points of T^2, etc.), and these are the only fixed points because T has integral coefficients.

The action of the cat map is illustrated in Fig. 142. After just one iteration the cat is wound around the torus in complicated filaments; its dissociation arises from the hyperbolic nature of T which causes initially close points to map far apart.

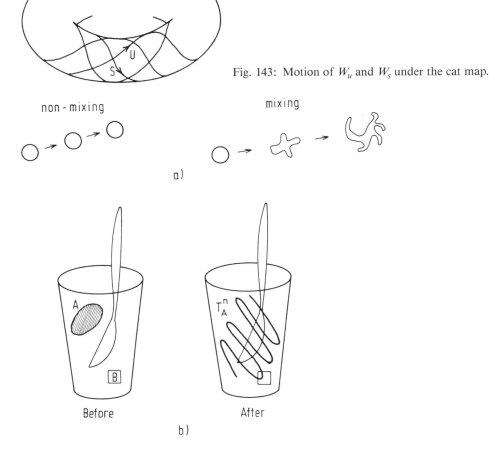

Fig. 143: Motion of W_u and W_s under the cat map.

non-mixing

mixing

a)

Before   After

b)

Fig. 144: a) Behavior of a volume element for nonmixing and for mixing transformations. b) Mixing of a drop of ink in a glass of water. (After Arnold and Avez, 1968.)

The axes of stretch (W_u) and compression (W_s) from (0, 0) lie along irrational directions and so wrap densely around the torus, never intersecting themselves but intersecting one another infinitely often, as shown in Fig. 143.

Since any set of iterates (which starts from a point (x_0, y_0) with x_0/y_0 = irrational) eventually covers the torus, "time" averages over the iterates are equal to "space" averages over the torus, and the motion generated by the cat map is ergodic.

However, the cat map has even a stronger property — mixing. In other words, the map distorts any area element so strongly, that it is eventually spread over the whole torus, just as a drop of ink (its volume corresponds to an area element in the cat map) is homogeneously distributed throughout a glass of water after it has been stirred (see Fig. 144).

Hierarchy of Classical Chaos

Table 13 gives an overview of the hierarchy of properties which indicates increasingly chaotic motion.

Table 13: Hierarchy of classical chaos.

Property	Definition	Example
Recurrent	The trajectory returns to a given neighborhood of a point an infinite number of times	Any Hamiltonian system (or area-preserving map) which maps a finite region of phase space onto itself
Ergodic	Time averages can be replaced by averages over phase space ↔ Zero is a simple eigenvalue of the Liouville operator L.	$x_{n+1} = x_n + b \bmod 1$ b = irrational
Mixing	Correlation functions decay to zero in the infinite time limit → L has one simple eigenvalue 0 and the rest of the spectrum is continuous.	Cat map
K-system	The map has a positive K-entropy, i.e. close orbits separate exponentially → L has a Lebesque spectrum with denumerably infinite multiplicity.	Cat map

The first entry in Table 13 contains the well-known Poincaré-recurrence theorem for Hamiltonian systems. It is simply a consquence of area-preserving motion in a finite region. We can draw an anology to what happens if we take a walk in a snow-covered finite square: eventually the area will be covered with footprints; and after some time, one is forced to walk on one's own prints (again and again).

Recurrence does not imply ergodicity because the allowed areas need not to be connected (there could be two squares). If the phase space is divided, the trajectory is confined to the region in which is started and does not cover the whole phase space.

More formally, a map f is called mixing, if

$$\lim_{n \to \infty} \rho\,[f^n(A) \cap B] = \rho(A)\rho(B) \tag{7.37}$$

for every pair of measurable sets A and B. Here ρ is the invariant measure of f. We used the abbreviation

$$\rho(A) \equiv \int_A dx\,\rho(x) \tag{7.38}$$

and assumed that the measure of the allowed phase space Γ, on which f acts, is normalized to unity, i.e. $\int_\Gamma dx\rho(x) = 1$.

If A and B correspond to the same point, eq. (7.37) reduces to

$$\lim_{n \to \infty} \int_\Gamma dx\,\rho(x)\,f^n(x)\,x \equiv \langle x_n x_0 \rangle = \left[\int_\Gamma dx\,\rho(x)\,x\right]^2 = \langle x_0 \rangle^2 \tag{7.39}$$

i.e. *mixing means that the autocorrelation function* $\langle(x_n - \langle x_0 \rangle)(x_0 - \langle x_0 \rangle)\rangle = \langle x_n x_0 \rangle - \langle x_0 \rangle^2$ *decays to zero* and "the system relaxes to thermal equilibrium". (The general proof can be found in the book by Arnold and Avez (1968) who actually show that a system is mixing if, and only if, $\lim_{n \to \infty}\langle F^*\,[f^n(x)]\,G(x)\rangle = \langle F^*(x)\rangle\,\langle G(x)\rangle$ for any square integrable complex-valued functions F and G.)

Although ergodicity (of course) implies recurrence, it does not imply mixing. Consider, for example, the map

$$x_{n+1} = x_n + b \bmod 1 \equiv f(x_n) \tag{7.40}$$

which shifts a point x_0 on a unit circle by b.

The map is ergodic for irrational values of b, because then a given starting point x_0 never returns to itself, as it does for rational $b = p/q$, (p, q integers) after q steps, and the images of x_0 cover the circle uniformly. The Liapunov exponent for this map is

$$\lambda = \lim_{n \to \infty} \frac{1}{n} \log \left|\frac{dx_n}{dx_0}\right| = 0 \tag{7.41}$$

i.e. (7.40) is an example that shows a) ergodicity without sensitive dependence on the initial conditions, and b) ergodicity without mixing. The last statement follows because the overlap of the images $f^n(A)$ of a line element A with another (line) element B is either finite or zero (according to the number of iterations) and never reaches a finite equilibrium value as required by eq. (7.37) (see Fig. 145). (Note that for simplicity we have replaced in this example "area" elements by "line" elements.)

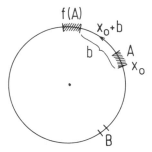

Fig. 145: Translations on a circle show ergodicity but they are not mixing.

Typical systems that show mixing are the cat map (Fig. 142) and the baker's transformation (Fig. 70a). In both cases a given volume element becomes distorted into finer and finer filaments that eventually cover the whole phase space uniformly. But, the rate at which volume elements become stretched need not be exponential (as in the examples quoted above), i. e. a system that shows mixing need not to be a K-system. These examples show that the properties in Table 12 indeed characterize increasingly chaotic motion.

We have also indicated in this table that the hierarchy of classical chaos is mirrored by the spectrum of eigenvalues of the Liouville operator. Let us briefly explain this fundamental relation which allows a characterization of classical chaos without considering individual trajectories.

The Liouville operator L determines the time evolution of the density $\rho\,(\vec{p}, \vec{q})$ in phase space:

$$\frac{\mathrm{d}}{\mathrm{d}t}\,\rho\,(\vec{p}, \vec{q}) = \dot{\vec{q}}\,\frac{\partial \rho}{\partial g\,\dot{\vec{q}}} + \dot{\vec{p}}\,\frac{\partial \rho}{\partial \dot{\vec{p}}} = \tag{7.42}$$

$$= \left[\frac{\partial H}{\partial \vec{p}}\,\frac{\partial}{\partial \vec{q}} - \frac{\partial H}{\partial \vec{q}}\,\frac{\partial}{\partial \vec{p}}\right]\rho \equiv -i\mathrm{L}\rho \tag{7.43}$$

$$\rightarrow \rho\,(t) = \mathrm{e}^{-it\mathrm{L}}\,\rho\,(0)\;. \tag{7.44}$$

Here we used Hamiltons's equations, and (7.43) defines L. It is useful to introduce the eigenvalues λ of L via

$$\mathrm{e}^{-i\mathrm{L}}\,\varphi\,(\vec{x}) = \mathrm{e}^{i\lambda}\,\varphi\,(\vec{x})\;;\quad \vec{x} = (\vec{p}, \vec{q}) \tag{7.45}$$

where $\varphi\,(\vec{x})$ is a complex, square integrable function in phase space. According to Table 13, different degrees of classical chaos correspond to different spectra of λ (the arrows indicate the direction of the statement). We explain this correspondence by two examples and refer to the cited literature for the general proofs.

First we consider two uncoupled harmonic oscillators whose Hamiltonion reads in action-angle variables:

$$H_{osc} = \omega_1 J_1 + \omega_2 J_2 \tag{7.46}$$

where ω_1, ω_2 are the oscillator frequencies. Eqns. (7.43 – 45) then become

$$-iL_{osc} \rho = \left[\omega_1 \frac{\partial}{\partial \theta_1} + \frac{\partial}{\partial \theta_2} \right] \rho \tag{7.47}$$

$$e^{-iL_{osc}} \varphi(\theta_1, \theta_2) = e^{i\lambda} \varphi(\theta_1, \theta_2) \tag{7.48}$$

where φ is periodic in the angles θ_1 and θ_2. These equations have the obvious solutions

$$\varphi(\theta_1, \theta_2) \propto e^{2\pi i(n_1 \theta_1 + n_2 \theta_2)} \tag{7.49}$$

$$\rightarrow \lambda = 2\pi(n_1 \omega_1 + n_2 \omega_2) \tag{7.50}$$

where n_1 and n_2 are integers.

The motion of the two oscillators on the tours (see Fig. 127) is only ergodic if ω_1/ω_2 is irrational, i.e. $\lambda \propto n_1 \omega_1 + n_2 \omega_2 = 0$ only for $n_1 = n_2 = 0$, and $\lambda = 0$ is a simple eigenvalue. For nonergodic motion ω_1/ω_2 = rational, and $\lambda = 0$ is degenerate. It is quite plausible that ergodicity and a nondegenerate eigenvalue $\lambda = 0$ correspond to each other because only then the equation for the time invariant density ρ,

$$e^{-iL} \rho = \rho \tag{7.51}$$

has a unique solution.

Eq. (7.44) can be extended to maps $\vec{x}_{n+1} = \vec{G}(\vec{x}_n)$ by

$$e^{-iL} \varphi(\vec{x}) \equiv \varphi[\vec{G}^{-1}(\vec{x})] = e^{i\lambda} \varphi(\vec{x}) . \tag{7.52}$$

As a further example, we consider the cat map (7.35) which acts on a torus so that we can expand φ as

$$\varphi(\vec{x}) = \sum_{\vec{m}} e^{2\pi i \vec{m} \cdot \vec{x}} \tilde{\varphi}(\vec{m}) \tag{7.53}$$

where $\vec{m} = (m_1, m_2)$; m_1, m_2 integers. Using the fact that the transformation matrix T is symmetric, we obtain from (7.52–53) after straightforward manipulations

$$\tilde{\varphi}(T\vec{m}) = e^{i\lambda} \tilde{\varphi}(\vec{m}) . \tag{7.54}$$

The point $\vec{m} = 0$ yields the only fixed point in (7.54), i.e. $\lambda = 0$ is again a simple eigenvalue that corresponds to a constant invariant density. The action of T on the other $\vec{m}$-values is explained in Fig. 146. If we relabel the $\vec{m}$'s according to their hyperbolas (a) and their place on it (j), i.e. $\vec{m} \triangleq (a, j)$, then eq. (6.54) can be written as

$$e^{-iL} \tilde{\varphi}(a,j) \equiv \tilde{\varphi}(a, j + 1) = e^{i\lambda} \tilde{\varphi}(a,j) \tag{7.55}$$

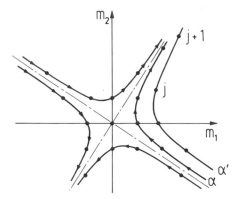

Fig. 146: Except for the origin, all points is the $\vec{m}$-plane are mapped under the action of the matrix T along hyperbolas because the eigenvalues of T are $\lambda_1 = (3 + \sqrt{5})/2 > 1$ and $\lambda_2 = 1/\lambda_2 < 1$.

ie. e^{-iL} is a translation operator in the variable j. The corresponding spectrum of L is continuous (note that the j's are not limited) and denumerably infinite degenerate (via the a's). A spectrum λ, which contains every real number with the same multiplicity and for which the spectral weight is just $d\lambda$, is called the Lebesque spectrum. The cat map is an example for a K-system. These systems have (also in general) Lebesque spectra with denumerably infinite multiplicity.

Three Classical K-Systems

Let us now present some physical examples of K-systems that exhibit ergodic and mixing behavior.

First, we consider the famous hard-sphere fluid whose mixing was rigorously established by Sinai (1970). Because of the infinite contact potential, this is clearly not a perturbation to a simple system (e.g. of noninteracting particles). Fig. 147a shows how exponential separation of the trajectories results from collisions between the spheres' convex surfaces. It is worth emphasizing that Sinai's proof is valid for *two* discs moving on a torus, i.e. it does *not* require the thermodynamic limit of infinitely many particles.

Another system, which has only a few degrees of freedom, but which nevertheless exhibits ergodicity and mixing, is a free particle in a stadium, as shown in Fig. 147b. The exponential separation of trajectories is generated by the particular form of the boundary (Bunimovich, 1979).

7   Regular and Irregular Motion in Conservative Systems

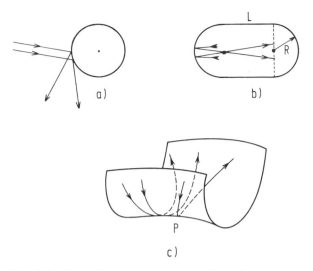

Fig. 147: Separation of trajectories for three chaotic systems: a) Sinai's billiards, b) a free particle in a stadium, c) a free particle on a surface with negative curvature.

Finally, we mention that the geodesic motion of a mass point on a compact surface with *overall* negative Gaussian curvature is also mixing and ergodic (Anosov, 1969). It can be already seen from the saddle-shaped surface in Fig. 147c (which has a negative curvature at *one* point P) how nearby trajectories separate along geodesics.

8 Chaos in Quantum Systems?

The existence of chaotic motion in classical conservative systems naturally leads to the question of how this irregularity manifests itself in the corresponding quantum systems. In a broader context, one might inquire about the nature of the solutions to the wave equations that arise, e. g., in plasma physics, optics, or acoustics, whose ray trajectories (WKB solution, geometric optics) are stochastic.

The question about the behavior of quantum systems whose classical limit exhibits chaos has been posed since the early days of quantum mechanics (Einstein, 1917) because it raises the problem of how to quantize a system which executes nonperiodic motion (at that time, periodic system were quantized via the Bohr-Sommerfeld quantization rule $\oint p \, dq = nh$, where h is Planck's constant). Since the discovery and the establishment of wave mechanics, we know how to proceed if we wish to learn about the time evolution of any quantum system: solve the time-dependent Schrödinger equation

$$\hat{H} \Psi = - \frac{h}{i} \frac{\partial}{\partial t} \Psi \tag{8.1}$$

where $\hat{H}$ is the Hamilton operator of the system, Ψ is its wave function, and $\hbar = h/2\pi$.

In order to develop some intuition for the changes which will arise if we pass from a classically chaotic system to its quantum mechanical version, we recall several major differences between classical and quantum systems:

— In contrast to classical mechanics (where a statistical description is only necessary if the system becomes chaotic in time), quantum mechanics allows *only* statistical predictions. Although the Schrödinger equation is linear in Ψ and can, e. g., be solved exactly for a harmonic oscillator with the result that Ψ depends regularly on time (i. e., there is no chaotic time behavior), this does *not* mean that the motion is completely deterministic, since $|\Psi(\vec{x}, t)|^2$ is only the probability density to find an electron at a space-time point $(\vec{x}, t)$.

— Because of Heisenberg's uncertaintly principle

$$\Delta p \, \Delta q > \hbar/2 \tag{8.2}$$

there are no trajectories in quantum mechanics (if one measures q with precision Δq, one disturbs the momentum p by Δp according to (8.2)). Therefore, the characterization of chaos based on the exponentially fast separation of nearby trajectories becomes useless for quantum systems.

— The uncertainty principle (8.2) implies also that points in 2 n-dimensional phase space within a volume $\hbar^n$ cannot be distinguished, i.e. the phase space becomes coarse grained. This means that regions in phase space in which the motion is classically chaotic (see Fig. 139), but which have volumes smaller than $\hbar^n$, are not "seen" in quantum mechanics; and for the corresponding quantum system, we expect a regular behavior in time. Thus the finite value of Planck's constant tends to suppress chaos. On the other hand, the limit $h \to 0$ becomes difficult (for quantum systems which have a classical counterpart which displays chaos) because if h becomes smaller, more and more irregular structures will appear.

In the following, we *distinguish* between (time-independent) *stationary Hamiltonians* and *time-dependent Hamiltonians,* which appear, for example, in the quantum version of the kicked rotator.

For systems with stationary Hamiltonians $\hat{H}$, the Schrödinger equation (8.1) can be reduced (with $\Psi = \Psi_0 \exp(-iEt/\hbar)$ to a linear eigenvalue problem for the energy levels E:

$$\hat{H} \Psi_0 = E \Psi_0 . \tag{8.3}$$

As long as the levels are discrete, Ψ behaves regularly in time and there is no chaos. But, there remain the fundamental questions: under what circumstances will this be the case and whether there are still differences between the energy spectra of a quantum system with a regular classical limit and a quantum system whose classical version displays chaos?

Information about the behavior of systems with time-dependent Hamiltonians are, for example, relevant for the problem of how energy is distributed in the energy ladder of a molecule excited by a laser beam, i.e. they are related to the practical problem of laser photochemistry.

More specifically, the answers to the following questions are sought: Does quantum chaos exist? How can one characterize it? Is there an equivalent to the hierarchy shown in Table 12 in quantum mechanics? What happens to the KAM-theorem for quantized motion, etc.? Up to now there are more questions than answers.

To get at least some insight into these problems, we consider several model systems.

In Section 8.1 we investigate the quantized version of the cat map (whose classical motion is purely chaotic) and show that it displays no chaos because the finite value of Planck's constant, together with the doubly periodic boundary conditions, restrict the eigenvalues of the time-evolution operator to a discrete set, such that the motion becomes completely periodic.

In the subsequent section we describe a calculation by McDonald and Kaufmann (1979), which shows that the energy spectrum of a free quantum particle in a stadium (for which the classical motion is chaotic) differs drastically from that of a free (quantum) particle in a circle (for which the classical motion is regular).

Finally, in the last section we demonstrate (by mapping the system to an electron localization problem) that a kicked quantum rotator shows no diffusion, whereas its classical counterpart displays deterministic diffusion above a certain threshold.

8.1   The Quantum Cat Map

To see how a conservative system, which classically behaves completely chaotically, changes its behavior for nonzero values of Planck's constant, we quantize a modification of Arnold's cat map. (The familiar cat map (7.35) cannot be quantized because the corresponding time-evolution operator does not preserve the periodicity of the wave function of the torus, see Hannay and Berry, 1980.)

Let us recall that the allowed phase space of a classical cat map is the unit torus. In this example, the phase points develop according to the dynamical law

$$\begin{pmatrix} p_{n+1} \\ q_{n+1} \end{pmatrix} = \begin{pmatrix} 1 & 2 \\ 2 & 3 \end{pmatrix} \begin{pmatrix} p_n \\ q_n \end{pmatrix}. \tag{8.4}$$

In quantum mechanics, eq. (8.4) becomes the Heisenberg equation of motion for the coordinate and momentum operators $\hat{q}_n$, $\hat{p}_n$ at time n, and the restriction of the classical phase space to a torus implies periodic boundary conditions for the quantum-mechanical wave function in coordinate *and* momentum space. In other words, the eigenvalues of *both* operators $\hat{p}$ *and* $\hat{q}$ only have discrete values which cover the torus by a lattice of allowed phase points, as shown in Fig. 148.

We will now show that the unit cell of this lattice is a square with a lattice constant, which is just Planck's quantum of the action h.

If the eigenvalues of $\hat{q}$ have a spacing $\Delta q = \dfrac{1}{N}$, i.e.

$$q = 0, \frac{1}{N}, \dots 1 \quad \text{where } N = \text{integer} \tag{8.5}$$

this implies (via the double periodicity of the wave function) the maximum momentum eigenvalue

$$p_{\max} = \hbar 2\pi / \left(\frac{1}{N}\right) = Nh \tag{8.6}$$

and a spacing $\Delta p = h$, i.e. the eigenvalues of $\hat{p}$ are

$$p = 0, h, 2h, \dots, Nh . \tag{8.7}$$

Because the allowed phase space has unit area, we have

$$1 = q_{max}p_{max} = Nh \tag{8.8}$$

$$\text{i.e.} \quad h = \frac{1}{N} \rightarrow \Delta p = \Delta q = h. \tag{8.9}$$

This requirement makes the quantum version of the cat map somewhat unrealistic. But if we assume for a moment that Planck's constant h is a free parameter and the quantum case is only defined by $h \neq 0$ such that eq. (8.9) makes sense, then it follows from (8.5) and (8.7) that in quantum mechanics only phase points with a rational ratio p/q are allowed. This means that the points with irrational ratios, which were the only ones in the classical cat maps which lead to chaotic trajectories, are forbidden in quantum mechanis. It is therefore reasonable to expect that the quantum version of the cat map will not exhibit chaos.

It has indeed been found by Hannay and Berry (1980) that the time-evolution operator $\hat{U}$ for the quantum cat map is periodic (i. e. for every N there exists an n (N) such that $\hat{U}^n = \hat{1}$) and has a discrete spectrum of eigenvalues. This implies that all expectation values for the cat map are periodic in time. In other words, the finite values of Planck's constant and the doubly periodic boundary conditions restrict the eigenvalues (of the time-evolution operator) in the quantum version of Arnold's cat map such that chaotic motion becomes impossible.

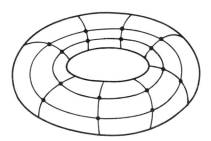

Fig. 148: Allowed phase points for the quantized version of a cat map (schematic).

8.2 A Quantum Particle in a Stadium

Although we have seen in the previous section that a quantum system with a chaotic classical limit does not necessarily also behave chaotically, one nevertheless expects some difference between a quantum system having a classical counterpart, which shows irregular motion, and a quantum version of an integrable classical system having regular trajectories. To cast some light on this problem, McDonald and Kaufmann (1979) calculated numerically the wave functions and spectra of a free particle in a stadium and in a circular disc by solving the Schrödinger equation for a free particle in two dimensions

$$\vec{\nabla}^2 \psi = E\psi \tag{8.10}$$

with the boundary condition $\psi(x, y) = 0$ at the "walls".

Their results are summarized in Fig. 149:

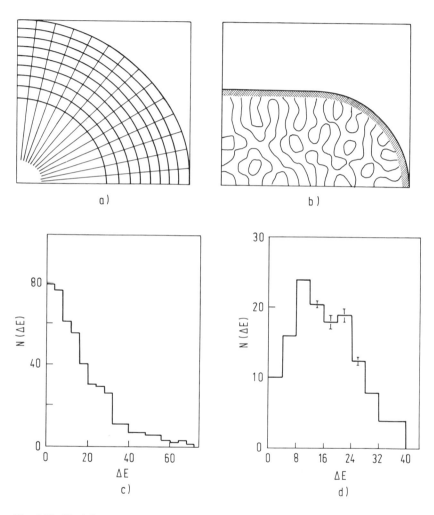

Fig. 149: Nodal curves [$\psi(x, y) = 0$] for one quadrant of the (odd-odd parity) eigenfunctions in a disc (a) and in a stadium (with dimensions $R = a$) (b). Distribution $N(\Delta E)$ of (odd-odd parity) energy level spacings for a circular boundary (c) and for a stadium boundary (d). (After McDonald and Kaufmann, 1979.) Note that $\Delta E = E_{j+1} - E_j$ is the spacing between neighboring levels, j increases with energy.

a) The eigenfunctions of the stadium problem show irregular nodal curves (where $\psi(x, y) = 0$) in contrast to the regular curves for the circle.

b) The distribution $N(\Delta E)$ of the eigenvalue spacings ΔE for the circle shows a maximum at $\Delta E = 0$, i. e. there is a high probability of level degeneracies, and one finds *level clustering*. It has been proved by Berry and Tabor (1977) that for integrable systems $N(\Delta E) \propto \exp(-\Delta E \cdot \text{const})$. (An exception is a quantum mechanical oscillator for which $N(\Delta E)$ is a delta function at $\Delta E = \hbar\omega_0$.) For the stadium $N(\Delta E)$ has a maximum at $\Delta E \neq 0$, i. e. there is *level repulsion*.

This level repulsion has also been found for the quantum version of Sinai's billiard (Berry, 1983, O. Bohigas et al., 1984), and it seems to be a characteristic feature of quantum system whose classical limit shows chaos. It is related to the fact that no symmetries exist in these systems, i. e. there are no degeneracies (and no selection rules which prevent mutual interaction of the levels) such that $\lim_{\Delta E \to 0} N(\Delta E) = 0$. Several theoretical explanations for this phenomenon have been offered, and an interesting connection to random matrix theory (which is used to explain level repulsion in nuclear spectra) has been suggested (Zaslavski, 1981, Berry, 1983, Bohigas et al., 1984). Note that the distribution of level spacings is related to the eigenvalue spectrum of the quantum version of the Liouville operator $\hat{L}$ because $\hat{L}|n\rangle \propto \langle m| \propto [\hat{H}, |n\rangle\langle m|] = (E_n - E_m)|n\rangle\langle m|$, where $\hat{H}$ is the Hamilton operator, and $|n\rangle$, $|m\rangle$ are its eigenfunctions.

8.3 The Kicked Quantum Rotator

We have already seen in Chapter 2 that deterministic diffusion serves as an indicator of chaos. It is, therefore, interesting to see whether this phenomenon also exists in quantum systems. (If the answer is yes, then we know that there is chaos in the quantum system). We show first that a classical kicked rotator, without damping, displays (for strong enough kicking forces) deterministic diffusion, and investigate subsequently its quantum version.

According to eq. (1.26), the equations of motion for the angle θ and the angular momentum p of a classical kicked rotator are

$$p_{n+1} = p_n - V'(\theta_n) \qquad n = 0, 1, 2 \ldots \tag{8.11a}$$

$$\theta_{n+1} = \theta_n + p_{n+1} = \theta_n - V'(\theta_n) + p_n \tag{8.11b}$$

where $V(\theta) = V(\theta + 2\pi)$ is the potential function of the kicking force.

Summation of (8.11a) over n yields

$$\langle(p_{n+1} - p_0)^2\rangle = \sum_{i,j}^{n} \langle V'(\theta_i) V'(\theta_j)\rangle \tag{8.12}$$

where $\langle \dots \rangle$ denotes the average over all initial points θ_0. If the correlations between the $V'(\theta_i)$ are short ranged (with range n_0), eq. (8.12) becomes

$$\langle (p_{n+1} - p_0)^2 \rangle = n \sum_j^{n_0} \langle V'(\theta_j) V'(\theta_0) \rangle \propto n \quad \text{for} \quad n \gg 1 \qquad (8.13)$$

i.e. the angular momentum of the kicked rotator diffuses.

It has, for example, been found numerically that a kicking potential of the form $V(\theta) = K \cdot \cos \theta$ generates deterministic diffusion (of the angular momentum) above a threshold $K_c \simeq 0.972$ (see Fig. 150).

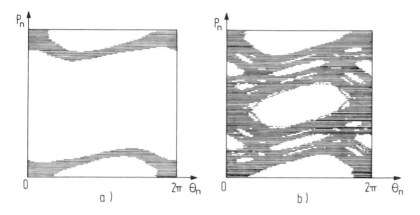

Fig. 150: A phase portrait of a classical kicked rotator with a potential function $K \cos \theta$, obtained by iterating eq. (7.11) and plotting successive points. a) For $K = 0.96$ different orbits in the shaded regions are still separated. b) For $K = 1.13$ the islands overlap and the angular momentum can diffuse. (After Chirikov, 1979.)

Another example is the "open cat map" in which the restriction of periodicity of the p_n is lifted. This can be viewed as a kicked rotator with a potential function $V(\theta) = -(K/2) \cdot (\theta \bmod 2\pi)^2$ and has the equations of motion

$$p_{n+1} = p_n + K\theta_n \qquad (8.14a)$$

$$\theta_{n+1} = \theta_n(1 + K) + p_n \qquad (8.14b)$$

where θ_n is always modulo 2π. Including the modulo restriction, eq. (8.14b) appears (apart from p_n, which does not seriously disturb our argument, and after division by 2π, which changes mod 2π to mod 1) similar to the map (2.1) that produced the Bernoulli shift. This means that for $K > 0$, eq. (8.14b) generates chaotic motion of the angles which leads via (8.12) to deterministic diffusion.

We now show that the quantum version of the kicked rotator does not diffuse. Instead, one finds either quantum resonance, i.e. the square of the angular momentum increases quadratically in time, or almost periodicity thus, the angular momentum is limited and recurs repeatedly arbitrarily close to its original value.

To unterstand this result, we use the idea of Fishman, Grempel, and Prange (1982) and map the kicked quantum rotator into a one-dimensional electron localization problem (about which several results are known). (The following derivation is due to V. Emery (private communication).) The time-dependent Hamiltonian of a kicked rotator can be written as

$$
\hat{H} = \begin{cases} \dfrac{\hat{V}(\theta)}{1 - \gamma} & \text{for} \quad 0 < t < 1 - \gamma \\[2em] \dfrac{\hat{T}}{\gamma} & \text{for} \quad 1 - \gamma < t < 1, \quad \text{with } \hat{T} = -\tau \dfrac{\partial^2}{\partial \theta^2} \end{cases}
\tag{8.15}
$$

where we have ignored the kinetic energy $\hat{T}$ during the delta kick which corresponds to the limit $\gamma \to 1$ in (8.15).

The time-evolution operator from time $t = n$ to time $t = n + 1$, i.e. before and after one kick, therefore becomes

$$
\hat{U} = e^{-i\hat{T}} e^{-i\hat{V}}
\tag{8.16}
$$

and its eigenstates $|\psi_\lambda\rangle$ are determined by

$$
\hat{U} |\psi_\lambda\rangle = e^{-i\lambda} |\psi_\lambda\rangle
\tag{8.17}
$$

where λ is the eigenvalue. This equation governs the time dependence of any state $|\varphi\rangle$ that develops with $\hat{U}$, because

$$
|\varphi(n)\rangle = \hat{U}^n |\varphi\rangle = \sum_\lambda e^{-in\lambda} c_\lambda |\psi_\lambda\rangle; \quad c_\lambda = \langle \psi_\lambda | \varphi \rangle .
\tag{8.18}
$$

We now rewrite (8.17) in form of a Schrödinger equation for an electron in a one-dimensional random chain. By using the explicit expression (8.16) for $\hat{U}$, (8.17) reads

$$
e^{-i\hat{T}} e^{-i\hat{V}} |\psi_\lambda\rangle = e^{-i\lambda} |\psi_\lambda\rangle
\tag{8.19}
$$

which for $\hat{E} \equiv \lambda \hat{I} - \hat{T}$ becomes

$$
e^{i\hat{E}} e^{-i\hat{V}} |\psi_\lambda\rangle = |\psi_\lambda\rangle .
\tag{8.20}
$$

With $|\psi_\lambda\rangle \equiv e^{i(\hat{V}/2)} |\omega\rangle$ this can be rewritten as:

$$
e^{i(\hat{V}/2)} |\omega\rangle - e^{i\hat{E}} e^{-i(\hat{V}/2)} |\omega\rangle = 0
\tag{8.21}
$$

or

$$\left[(1 - e^{i\hat{E}}) \cos \frac{\hat{V}}{2} + i(1 + e^{i\hat{E}}) \sin \frac{\hat{V}}{2} \right] |\omega\rangle = 0 \qquad (8.22)$$

from which we obtain

$$i(1 + e^{i\hat{E}}) \left[\frac{1}{i} \frac{1 - e^{i\hat{E}}}{1 + e^{i\hat{E}}} + \frac{\sin(\hat{V}/2)}{\cos(\hat{V}/2)} \right] \cos \left(\frac{\hat{V}}{2} \right) |\omega\rangle = 0 . \qquad (8.23)$$

We, therefore, have to find the solutions of

$$\left[\tan \frac{\hat{E}}{2} - \tan \frac{\hat{V}}{2} \right] |u\rangle = 0, \quad \text{when} \quad |u\rangle = \cos \frac{\hat{V}}{2} |\omega\rangle . \qquad (8.24)$$

The periodic boundary conditions $\psi_\lambda (\theta + 2\pi) = \psi_\lambda (\theta)$ yield $u(\theta + 2\pi) = u(\theta)$, i. e. $u(\theta)$ can be expanded in a Fourier series:

$$u(\theta) = \sum_m u_m \, e^{im\theta} . \qquad (8.25)$$

Note that $e^{im\theta}$ is simply the eigenfunction of the angular momentum operator. Thus, (8.24) can be written as

$$T_m u_m + \sum_{r \neq 0} W_r u_{m+r} = \varepsilon u_m; \; \varepsilon = W_0 \qquad (8.26)$$

where

$$T_m \equiv \tan \left[\frac{1}{2} (\lambda - \tau m^2) \right] \quad \text{and} \quad W_r = \frac{1}{2\pi} \int_{-\pi}^{\pi} d\theta \, e^{ir\theta} \tan \left[\frac{\hat{V}(\theta)}{2} \right] .$$

Eq. (8.26) is the Schrödinger equation for an electron on a chain with on-site potentials T_m and hopping matrix elements W_r. The integer eigenvalues m of the angular momentum of the kicked rotator correspond to the lattice sites in the conduction problem.

Two cases must be distinguished:

a) For rational values of $\tau/(2\pi) = p/q$, where p and q both are mutually prime integers, the electrons described by (8.26) move freely in a periodic potential and are completely delocalized. For the rotator problem this means that its angular momentum is unbounded in time, i. e. all eigenvalues m can be achieved. In fact the square of the angular momentum increases quadratically in time. This phenomenon is termed quantum resonance and occurs for all rational values of $\tau/(2\pi)$. We will explain this phenomenon for the simplest case $p/q = 1$. The effect of the time-evolution operator on any periodic wave function ψ then becomes

$$\hat{U}\,|\,\psi\rangle \triangleq e^{2\pi i(\partial^2/\partial\theta^2)}\,e^{-iV(\theta)}\,\psi\,(\theta) = e^{-iV(\theta)}\,\psi\,(\theta) \tag{8.27}$$

since we can expand $e^{-iV}\psi$ in a Fourier series:

$$e^{-iV(\theta)}\,\psi\,(\theta) = \sum_{m=-\infty}^{\infty}\,A_m\,e^{im\theta} \tag{8.28}$$

and

$$(e^{-2\pi i(\partial^2/\partial\theta^2)})\,e^{im\theta} = e^{-2\pi im^2}\,e^{im\theta} = e^{im\theta}\ . \tag{8.29}$$

For the expectation value of the square of the angular momentum with any periodic wave function after n kicks, we therefore find:

$$\langle p^2\rangle \propto \langle\psi\,|\,(\hat{U}^+)^n\,\frac{\partial^2}{\partial\theta^2}\,\hat{U}^n\,|\,\psi\rangle \propto \int_{-\pi}^{\pi}\,d\theta\,\psi^*(\theta)\,e^{inV(\theta)}\,\frac{\partial^2}{\partial\theta^2}\,e^{-inV(\theta)}\,\psi\,(\theta)$$

$$\propto n^2\,\langle\psi\,|\,\left(\frac{\partial V}{\partial\theta}\right)^2\,|\,\psi\rangle + O\,(n)\ . \tag{8.30}$$

This quadratic increase in time is clearly a quantum effect because (8.27) holds only for integer values of m, i.e. for a quantized angular momentum.

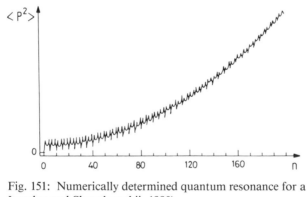

Fig. 151: Numerically determined quantum resonance for a kicked rotator at $\tau = 8\,\pi/5$ (After Izraelev and Shepelyanskii, 1980).

b) Next we consider the case where $\tau/(2\,\pi)$ is irrational. The potential $T_m = \tan[(\lambda - m^2\tau)/2]$ then becomes random instead of periodic because $[(\lambda - m^2\tau)/2]$ mod π behaves like a random number generator. (Note that $\tan x$ is periodic with period π and its argument can, after division by π, be written as $x_m = [\lambda/(2\pi) - m^2\tau/(\pi)]$ mod 1. If $\tau/(2\pi)$ is expressed in binary representation and one considers, for example, values $m^2 = 2^n$, then it is seen that the $x_{m=2^n}$ are generated by a Bernoulli shift from a irrational number and are, hence, truly random.)

Intuitively, one expects that an electron in a one-dimensional random potential has a strong tendency to localize since there is (in contrast to higher dimensions) only one way to move from one point to the next, and this could be easily blocked by a potential barrier. It is in fact well known from the work of Anderson (1958) and Ishii (1973) (but by no means trivial to prove) that all electrons in a one-dimensional random potential are localized (for short-ranged hopping matrix elements). The physical reason for this is that, in the one-dimensional case, the random potential changes the phase of the wave function at every site, and this random dephasing eventually leads to localization.

The electron is, therefore, confined to a finite range of m's, i.e. the angular momentum of the rotator is bounded and does not increase in time; in other words, there is no diffusion of momentum in contrast to the classical case. Fig. 152 shows the time dependence of the energy of a periodically kicked rotator, numerically calculated for an irrational value of $\tau/2\pi$. It can be seen that the oscillations in energy are not only bounded but recur many times.

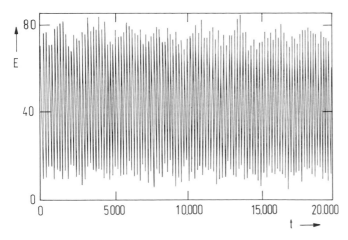

Fig. 152: Numerical result for the expectation value of the energy $E \propto \langle p^2 \rangle$ (of a kicked quantum rotator) as a function of the number n of pulses for an irrational value of $\tau/(2\pi)$ (after Hogg and Hubermann, 1982).

It has been proved by Hogg and Hubermann (1982) that if the wave function can be normalized (i.e., if we know that the angular momentum does not diffuse) then both the wave function and the energy return arbitrarily close to their initial values arbitrarily often. This time dependence is called *almost periodic* in contrast to the quasiperiodic motion mentioned in Chapter 6. (For almost periodic functions $f(t)$ there exists a relatively dense set $\{\tau_\varepsilon\}$ such that $| f(t + \tau_\varepsilon) - f(t) | < \varepsilon$ for any $\varepsilon > 0$. $\{\tau_\varepsilon\}$ is relatively dense if there exists a T_ε such that each interval of length T_ε on the real axis contains at least one τ_ε.)

We have seen, above, that up to now no quantum system seems to exist which exhibits deterministic chaos (indicated either by a continuous power spectrum or deterministic diffusion). Nevertheless, there is a difference in the behavior of quantum systems with a chaotic classical counterpart and those (quantum systems) with a regular classical limit.

Let us finally mention an interesting calculation of Gutzwiller (1983) for an electron which is scattered from a non-compact surface with negative curvature. It shows that the phase shift as a function of momentum is essentially given by the phase angles of the Riemann zeta function on the imaginary axis, at a distance 0.5 from the famous critical line. This phase shift displays features of chaos because it is able to mimick any given smooth function. It, therefore, seems that the chaotic nature of quantum systems which are described by *wave* mechanics is of a rather subtle and "softer" kind than the chaos in classical mechanics.

These comments indicate that the question of stochasticity in quantum mechanics is still far from being solved.

Outlook

In this book, we presented an introduction to deterministic chaos, stressing the importance of self-similar structures and renormalization-group ideas. Let us now take a glance at future possible developments by indicating several topics not dealt within previous chapters.

First of all, there is the problem of *chaotic motion in spatially coupled nonlinear systems,* such as: coupled heart cells, chemical reactions (Vidal and Pacault, 1984) where the diffusion term is included (i. e. $\dot{\vec{c}} = \vec{F}(\vec{c}) + k\,\vec{\nabla}^2\,\vec{c}$ instead of eq. (1.7)), and the Navier-Stokes equations (Ruelle, 1983). Some pertinent questions are: How do these nonlinear oscillators influence each other? Do they synchronize? Does there exist something like spatial chaos? What is the influence of spatial motion on temporal chaos? What are the dimensions of strange attractors if one approaches fully developed turbulence? ...

In addition to the problems mentioned in the previous chapter, another major area is the question of the chaotic behavior of *quantum systems with dissipation,* such as lasers or Josephson-junctions, etc. (Graham, 1984). It is also interesting to note that in *quantum systems with many particles* the question of chaos is related to the fundamental problem of the "arrow of time" (Misra and Prigogine, 1980).

This list is, of course, by no means complete. From the mathematical point of view, the *very nature of a random number* is still an issue of interest (De Long, 1970), and we have not discussed the possible role that the close coexistence of chaos and regular motion could play for the *formation of structures in biology* (Hess and Markus, 1984).

It should also be noted that cellular automata (i. e. discrete approximations to partial differential equations in which all variables, time, space, and the signal only take integer values) seem to become an important tool for answering many of the questions mentioned above. (Farmer et al., 1984; Wolfram, 1985; Frisch et al., 1986).

But the major conclusion should be clear: *Since nature is nonlinear, one has always to reckon with deterministic chaos.* This means, however, that prediction about the future development of the field of deterministic chaos are as difficult or short ranged as predictions of chaotic motion itself, i. e. there is (fortunately) much room for surprises. Interestingly enough, already about 100 years ago, James Clerk Maxwell (the founder of the theory of electromagnetism) wrote the following far-sighted remark about the predictability of nonlinear, i. e. unstable systems (quoted after Berry, 1978):

"If, therefore, those cultivators of the physical science from whom the intelligent public deduce their conception of the physicist . . . are led in pursuit of the arcana of science to the study of the singularities and instabilities, rather than the continuities and stabilities of things, the promotion of natural knowledge may tend to remove that prejudice in favor of determinism which seems to arise from assuming that the physical science of the future is a mere magnified image of that of the past."

Appendix

A   Derivation of the Lorenz Model

References: see Chapter 1

Here we present a rather short derivation of the Lorenz model that should provide the reader with a feeling for the approximations involved. For a more rigorous treatment, we refer the reader to the original articles by Saltzmann (1961) and Lorenz (1963) and the monograph by Chandrasekhar (1961).

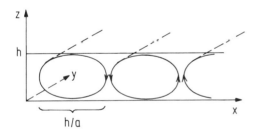

Fig. 153: Convection rolls and geometry in the Bénard experiment.

Consider the Rayleigh-Bénard experiment as depicted in Fig. 118. The liquid is described by a velocity field $\vec{v}(\vec{x}, t)$ and a temperature field $T(\vec{x}, t)$. The basic equations which describe our system are

a) the Navier-Stokes equations:

$$\rho \, \frac{\mathrm{d}\vec{v}}{\delta t} = \vec{F} - \vec{\nabla} p + \mu \vec{\nabla}^2 \vec{v} \qquad\qquad (A.1)$$

b) the equation for heat conduction:

$$\frac{\mathrm{d}T}{\mathrm{d}t} = \kappa \vec{\nabla}^2 T \qquad\qquad (A.2)$$

c) the continuity equation:

$$\frac{\partial \rho}{\partial t} + \mathrm{div}\,(\rho\,\vec{v}) = 0 \tag{A.3}$$

with the boundary conditions

$$T(x, y, z = 0, t) = T_0 + \Delta T$$
$$T(x, y, z = h, t) = T_0\,. \tag{A.4}$$

Here ρ is the density of the fluid, μ is its viscosity, p is the pressure, κ is the thermal conductivity, and $F = \rho g \vec{e}_z$ is the external force in the $\vec{e}_z$-direction due to gravity. The *fundamtental nonlinearity in hydrodynamics* comes from the term $\dot{\vec{v}} = (\vec{v} \cdot \vec{\nabla})\,\vec{v} + \partial\vec{v}/\partial t$ (which is quadratic in $\vec{v}$) in the Navier-Stokes equation (A.1).

To simplify the calculation, it is assumed a) that the system is translationally invariant in the y-direction so that convection rolls extend to infinity as shown in Fig. 153, and b) that the ΔT-dependence of all coefficients — except in $\rho = \bar{\rho}\,(1 - \alpha \Delta T)$ — can be neglected (Boussinesq approximation). The continuity equation thus becomes

$$\frac{\partial u}{\partial x} + \frac{\partial w}{\partial z} = 0 \quad \text{with} \quad u = v_x \quad \text{and} \quad w = v_z \tag{A.5}$$

and, it is, therefore, convenient to introduce a function $\psi\,(x, z, t)$ with

$$u = -\frac{\partial \psi}{\partial z} \quad \text{and} \quad w = \frac{\partial \psi}{\partial x} \tag{A.6}$$

such that (A.5) is automatically fulfilled.

As a next step we introduce the deviation $\theta\,(x, z, t)$ from the linear temperature profile via

$$T(x, z, t) = T_0 + \Delta T - \frac{\Delta T}{h}\,z + \theta\,(x, z, t)\,. \tag{A.7}$$

Using (A.6) and (A.7) the basic equations can, according to Saltzmann, be written as

$$\frac{\partial}{\partial t}\,\vec{\nabla}^2 \psi = -\frac{\partial\,(\psi, \vec{\nabla}^2 \psi)}{\partial\,(x, z)} + v\,\vec{\nabla}^4 \psi + g\alpha\frac{\partial \theta}{\partial x} \tag{A.8}$$

$$\frac{\partial}{\partial t}\,\theta = -\frac{\partial\,(\psi, \theta)}{\partial\,(x, z)} + \frac{\Delta T}{h}\,\frac{\partial \psi}{\partial x} + \kappa\,\vec{\nabla}^2 \theta \tag{A.9}$$

where

$$\frac{\partial(a, b)}{\partial(x, z)} \equiv \frac{\partial a}{\partial x} \cdot \frac{\partial b}{\partial z} - \frac{\partial a}{\partial z} \cdot \frac{\partial b}{\partial x}, \tag{A.10}$$

$$\vec{\nabla}^4 \equiv \frac{\partial^4}{\partial x^4} + \frac{\partial^4}{\partial z^4}$$

$\nu \equiv \mu/\bar{\rho}$ is the kinematic viscosity, and the pressure term was eleminated by taking the curl in the Navier-Stokes equations.

In order to simplify (A.8) and (A.9), Lorenz used free boundary conditions:

$$\theta(0, 0, t) = \theta(0, h, t) = \psi(0, 0, t) = \psi(0, h, t)$$
$$= \vec{\nabla}^2 \psi(0, 0, t) = \vec{\nabla}^2 \psi(0, h, t) = 0 \tag{A.11}$$

and retained only the lowest order terms in the Fourier expansions of ψ and θ and proposed the following ansatz:

$$\frac{a}{1 + a^2} \frac{1}{\kappa} \psi = \sqrt{2}\, X(t) \sin\left(\frac{\pi a}{h} x\right) \sin\left(\frac{\pi}{h} z\right) \tag{A.12}$$

$$\frac{\pi R}{R_c \Delta T} \theta = \sqrt{2}\, Y(t) \cos\left(\frac{\pi a}{h} x\right) \sin\left(\frac{\pi}{h} z\right)$$

$$- Z(t) \sin\left(\frac{2\pi z}{h}\right) \tag{A.13}$$

where $R \equiv (g a h^3/\kappa \nu) \Delta T$ is the Rayleigh number, a is the aspect ratio (see Fig. 153) and $R_c \equiv \pi^4 a^{-2} (1 + a^2)^3$. By inserting (A.12–13) into (A.8–9) and neglecting higher harmonics, one obtains finally the Lorenz model:

$$\dot{X} = -\sigma X + \sigma Y \tag{A.14a}$$

$$\dot{Y} = -XZ + rX - Y \tag{A.14b}$$

$$\dot{Z} = XY - bZ \tag{A.14c}$$

where the dot denotes the derivative with respect to the normalized time $\tau \equiv \pi^2 h^{-2}(1 + a^2)\kappa t$; $\sigma \equiv \nu/\kappa$ is the Prandl number, $b \equiv 4(1 + a^2)^{-1}$, and $r = R/R_c \propto \Delta T$ is the external control parameter.

B   Stability Analysis and the Onset of Convection and Turbulence in the Lorenz Model

References: see Chapter 1

Let us write the Lorenz equations (A.14) in the compact form

$$\dot{\vec{X}} = \vec{F}(\vec{X}) \tag{B.1}$$

and linearize around the fixed points

$$\vec{X}_1 = \vec{0}; \quad \vec{X}_2 = (\pm \sqrt{b(r-1)}; \quad \pm \sqrt{b(r-1)}; \quad r-1) \tag{B.2}$$

which are determined by

$$\vec{F}(\vec{X}_{1,2}) = \vec{0}. \tag{B.3}$$

The first fixed point $\vec{X}_1 = \vec{0}$ corresponds to thermal conductivity without motion of the liquid, and its stability matrix

$$\frac{\partial F_i}{\partial X_j}\bigg|_{\vec{X}_1} = \begin{pmatrix} -\sigma & \sigma & 0 \\ r & -1 & 0 \\ 0 & 0 & -b \end{pmatrix} \tag{B.4}$$

has the eigenvalues

$$\lambda_{1,2} = -\frac{\sigma+1}{2} \pm \frac{1}{2}\sqrt{(\sigma+1)^2 + 4(r-1)\sigma}; \quad \lambda_3 = -b. \tag{B.5}$$

Thus, $\vec{X}_1 = \vec{0}$ is stable — i.e. all eigenvalues are negative — for $0 < r < 1$. The Bénard convection starts at $r = 1$ because then $\lambda_1 = 0$, and this is just where the second fixed point $\vec{X}_2$ (which corresponds to moving rolls, as shown in Fig. 154) takes over. The stability matrix for $\vec{X}_2$ is

$$\frac{\partial F_i}{\partial X_j}\bigg|_{\vec{X}_2} = \begin{pmatrix} -\sigma & \sigma & 0 \\ 1 & -1 & c \\ c & c & -b \end{pmatrix}; \quad c \equiv \pm \sqrt{b(r-1)}. \tag{B.6}$$

Its eigenvalues are the roots of the polynomial

$$P(\lambda) = \lambda^3 + (\sigma+b+1)\lambda^2 + b(\sigma+r)\lambda + 2b\sigma(r-1) = 0. \tag{B.7}$$

One sees immediately that for $r = 1$ we have $\lambda_1 = 0$, $\lambda_2 = -b$, and $\lambda_3 = -(\sigma + 1)$, i.e. the convection fixed point is marginally stable, and Fig. 154 shows that it is stable for $1 < r < r_1$. At $r_1 < r_c$ two of the eigenvalues become complex, i.e. two limit cycles result which are stable so long as the real part of the complex eigenvalues is smaller than zero. For $r = r_c$ these real parts become zero, i.e. we have two eigenvalues $\lambda = \pm i\lambda_0$, which lead via (B.7) to

$$r_c = \sigma \frac{\sigma + b + 3}{\sigma - b - 1} \left(= 24.7368 \quad \text{for} \quad \sigma = 10, b = \frac{8}{3} \right).$$

Above r_c the limit cycles become unstable (the complex eigenvalues have positive real parts), and chaos sets in. This analysis is consistent with the numerical result obtained by Lorenz, who found chaotic behavior for $\sigma = 10$, $b = 8/3$ above $r_c = 24.74$.

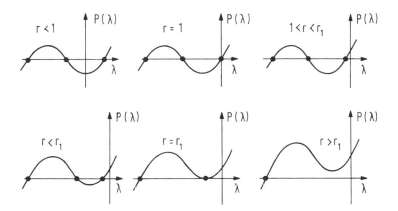

Fig. 154: Qualitative behavior of the polynomial $P(\lambda)$.

C   The Schwarzian Derivative

References: see Chapter 3

Not all unimodal functions (i.e. continuously differentiable maps f that map the unit interval $[0,1]$ onto itself with a single maximum at $x = 1/2$ and are monotonic for $0 \le x \le 1/2$ and $1/2 < x \le 1$) display an infinite sequence of pitchfork bifurcations.

In addition to being unimodular, the Schwarzian derivative of f

$$Sf \equiv \frac{f'''}{f'} - \frac{3}{2} \left(\frac{f''}{f'} \right)^2 \propto \frac{d^2}{dx^2} [f'(x)]^{-1/2} \tag{C.1}$$

must be negative over the whole interval $[0,1]$. This is, for example, true for the logistic map, since $f'''(x) = 0$.

To make this requirement, which at first sight appears unusual, more plausible, we note the important property that $Sf < 0$ implies a negative Schwarzian derivative for all iterates of f, i.e. $Sf^n < 0$. This can be verified by direct calculation. As a consequence, it is found that at a fixed point x_0 of f that just becomes unstable, i.e.

$$f'(x_0) = -1 \tag{C.2}$$

and

$$f^{2\prime}(x_0) = [f'(x_0)]^2 = 1 \tag{C.3}$$
$$f^{2\prime\prime}(x_0) = f''(x_0)\{[f'(x_0)]^2 + f'(x_0)\} = 0$$

the third derivative of $f^2(x_0)$ becomes negative for $Sf < 0$, and, near $x_0 = 0$, $f^2(x)$ behaves as shown in Fig. 155, which can lead to a pitchfork bifurcation. The same figure shows that a pitchfork bifurcation becomes impossible for $Sf > 0$.

The importance of the Schwarzian derivative had first been noted by Singer (1978), who showed that unimodal maps with $Sf < 0$ cannot have more than one periodic attractor. Later Guckenheimer and Misurewicz proved that, in this case, all points in $[0,1]$ (i.e. with the exception of a set of measure zero) become attracted to it. The proofs and references can be found in the monograph by Collet and Eckmann (1980).

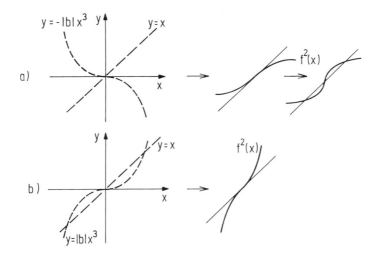

Fig. 155: Behavior of $f^2(x) = f^{2\prime}(x_0)x + bx^3$; $b \equiv f^{2\prime\prime\prime}(x_0)/3!$ near $x_0 = 0$ for a) $Sf < 0$, i.e. $b < 0$, and b) $Sf > 0$, i.e. $b > 0$.

D   Renormalization of the One-Dimensional Ising Model

References: see Chapter 3

The functional renormalization group which is used in this book has been constructed in analogy to the renormalization-group method for critical phenomena. This section explains the method for critical phenomena (which is simpler than the functional renormalization method) for the example of the one-dimensional Ising model. Although the one-dimensional Ising model has several strange features (its transition temperature is zero, etc., see below), these are outweighted by the fact that every renormalization-group step can be performed explicitly. It is assumed that the reader is familiar with the usual exact solution of this model that can be found in most textbooks on statistical mechanics.

The partition function of the one-dimensional Ising model has the well-known form

$$Z = \sum_{\{\sigma_i\}} e^{\sum_i \sigma_i \sigma_{i+1}} \qquad \text{(D.1)}$$

where $\beta = J/T$ is the ratio of coupling constant J and temperature T; the spin variables σ_i take the values $\sigma_i = \pm 1$, and the sites are $i = 0 \ldots N$.

The renormalizations-group steps are visualized in Fig. 156:

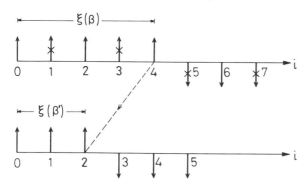

Fig. 156 Renormalization-group steps for the one-dimensional Ising Model. a) Spins with odd indices are integrated out. b) The correlation length in the renormalized system ($\beta \to \beta'$, $2i \to i$) becomes smaller.

First, we sum in (D.1) over all spin variables σ_i with odd i. Then we relabel the remaining variables with even i:

$$\sigma_{2i} \to \sigma_i/\alpha \qquad \text{(D.2)}$$

(for our simple example, we have $a = 1$; but already for the two-dimensional Ising model, one needs $a \neq 1$). Fig. 156 shows that the system of residual spins exhibits the same pattern as before and only two factors have changed: all lengths are reduced by a factor of two and the coupling between the residual spins becomes renormalized ($\beta \to \beta'$). At the transition temperature $T = T_c = 0$, the correlation length is infinite and the spin pattern is self-similar for all length scales, i.e. repeated applications of the renormalization-group procedure always lead to similar results.

To perform these steps explicitly, we consider a typical sum over an odd variable in (D.1):

$$Z_3 = \sum_{\sigma_3} e^{\beta(\sigma_2\sigma_3 + \sigma_3\sigma_4)} = 2\left[(\cosh\beta)^2 + \sigma_2\sigma_4(\sinh\beta)^2\right]. \tag{D.3}$$

This can be written as

$$Z_3 = c \cdot e^{\beta'\sigma_2\sigma_4} = c\left[\cosh\beta + \sigma_2\sigma_4\sinh\beta\right] \tag{D.4}$$

with

$$\th\beta' = (\th\beta)^2. \tag{D.5}$$

Eq. (D.5) is obtained by comparing the right-hand sides of (D.3) and (D.4), keeping in mind that σ_2 and σ_4 have only the values ± 1.

In the next step, we relabel the spins according to (D.2) and obtain the renormalized version of Z:

$$Z(\beta) = Z(\beta') = c^{N/2} \sum_{\{\sigma_i\}} e^{\beta' \sum_i^{N/2} \sigma_i\sigma_{i+1}}. \tag{D.6}$$

(The constant c will not be further considered because it cancels in all thermodynamic averages). The renormalized coupling β', between the residual spins, is according to (D.6):

$$\beta' = \text{Arth}\left[(\th\beta)^2\right] \equiv R_2(\beta). \tag{D.7}$$

Iteration of the renormalization procedure yields

$$\beta'' = R_2[R_2(\beta)] = R_4(\beta). \tag{D.8}$$

The last equal sign means that two repeated renormalizations are equivalent to one renormalization where only every fourth spin is retained, i.e. the renormalization-group operators R form a semigroup ("semi" because there exists no inverse element). The fixed points of (D.7) are

$$\beta^* = \infty \quad \text{and} \quad \beta^* = 0 \tag{D.9}$$

i. e. they occur at zero temperature (the transition temperature of the one-dimensional Ising model) and at infinite temperature. In both limits, the spin pattern is self-similar (the spin system is completely disordered at $T = \infty$, and at $T = 0$ all spins are aligned). For $\beta > 0$, the system is always driven (by repeated applications of R_2) to the stable fixed point $\beta^* = \infty$.

Because the correlation length ξ is reduced by a factor of two, after one renormalization step, we can immediately determine the temperature dependence of ξ via the following scaling argument:

$$\xi(\beta) = 2\xi(\beta') \tag{D.10}$$

$$\rightarrow \xi(\beta) = 2\xi\{\text{Arth}\,[(\text{th}\,\beta)^2]\} = 2^n\xi\{\text{Arth}\,[\text{th}\,\beta)^{2^n}]\} . \tag{D.11}$$

For $\beta \gg 1$, the variable n can be chosen such that

$$(\text{th}\,\beta)^{2^n} = \text{const.} \tag{D.12}$$

$$\rightarrow 2^n \propto 1/\log(\text{th}\,\beta) \tag{D.13}$$

$$\rightarrow \xi \propto 1/\log(\text{th}\,\beta) \tag{D.14}$$

This last relation can be verified by direct computation of the correlation function:

$$\langle \sigma_{j+r}\sigma_j \rangle \equiv \sum_{\{\sigma_i\}} e^{\beta \sum_i \sigma_i\sigma_{i+1}} \sigma_{j+r}\sigma_j \Big/ \Big(\sum_{\{\sigma_i\}} e^{\beta \sum_i \sigma_i\sigma_{i+1}} \Big) \tag{D.15}$$

$$= (\text{th}\,\beta)^r \equiv e^{-r/\xi} \tag{D.16}$$

where we used

$$\sum_{\sigma_{i+1}} e^{\beta\sigma_i\sigma_{i+1}} = 2\cosh\beta \tag{D.17}$$

and

$$\sum_{\sigma_{i+1}} e^{\beta\sigma_i\sigma_{i+1}} \sigma_{i+1} = 2\sigma_i \sinh\beta . \tag{D.18}$$

It should be noted that for more complicated systems (e. g. the two- or three-dimensional Ising model), the elimination of spin variables in one renormalization step leads to next nearest neighbor and higher order couplings (between the spins), and it is part of the art of renormalization to keep track of them.

E Decimation and Path Integrals for External Noise

References: see Chapter 3

Here we present a derivation of the scaling form of the Liapunov exponent (3.91) that follows an important article by Feigenbaum and Hasslacher (1982). Our main aim is to explain their decimation method which has, on the one hand, a wide range of potential applications (for example to the transition from quasi-periodicity to chaos in Chapter 6), and, on the other hand, close parallels to the renormalization of the one-dimensional Ising model (explained in Appendix D).

As a first step, we write the iterates of (3.87)

$$x_{n+1} = f(x_n) + \xi_n \tag{E.1}$$

as integrals over δ-functions:

$$x_1 = f(x_0) + \xi_0 = \int dx_1 x_1 \delta [x_1 - f(x_0) - \xi_0] \tag{E.2a}$$

$$x_2 = f[f(x_0) + \xi_0] + \xi_1 = \tag{E.2b}$$

$$= \int dx_1 dx_2 x_2 \delta [x_2 - f(x_1) - \xi_1] \delta [x_1 - f(x_0) - \xi_0]$$

$$\vdots$$

$$x_n = \int \prod_{j=1}^{n} dx_j x_n \delta [x_{j+1} - f(x_j) - \xi_j] \tag{E.2c}$$

The ξ_j are independent random variables with Gaussian probability distributions:

$$P_0\{\xi_j\} = \prod_j P[\xi_j; \sigma^2] \equiv \prod_j \frac{1}{\sqrt{2\pi}\sigma} e^{-\xi_j^2/2\sigma^2} . \tag{E.3}$$

If we use eqns. (E.2c), (E.3) and integrate over $\{\xi_i\}$, the average of x_n becomes

$$\langle x_n \rangle = \int \prod_j d\xi_j P_0\{\xi_j\} x_n = \tag{E.4}$$

$$= \int \prod_{j=1}^{n} dx_j x_n \prod_{i=0}^{n-1} P[x_{i+1} - f(x_i); \sigma^2] .$$

This average has the form of a path integral that is (if the time variable i is interpreted as a site index) reminiscent of the thermodynamic average of a magnet and, therefore, well-suited for a renormalization-group treatment. The idea is to perform the integration over the x_i, *step by step*, i.e. the renormalization-group treatment consists of integrating out all x_i with odd i's (this is called "decimation") and rescaling the variables such that the whole operation can be *repeated*.

Let us choose $n = 2^q$, q integer, and separate variables with even and odd indices in (E.4):

$$\langle x_n \rangle = \int \prod_1^{n/2} dx_{2i} \, x_n \prod_1^{n/2} dx_{2i-1} \prod_0^{n/2-1} P[x_{2i+2} - f(x_{2i+1}); \sigma^2] \times$$

$$\times \, P[x_{2i+1} - f(x_{2i}); \sigma^2] \, . \tag{E.5}$$

The relevant integrals over the odd variables,

$$I = \int dx_{2i+1} \exp \{-[x_{2i+2} - f(x_{2i+1})]^2/2\sigma^2 - [x_{2i+1} - f(x_{2i})]^2/2\sigma^2\}^2 \tag{E.6}$$

are evaluated using the saddle-point approximation that is valid for small noise amplitudes $\sigma \ll 1$.

If we have an integral over a function which is sharply peaked at x^*, the simplest form of the saddle-point approximation consists of replacing the integral by the integrand taken at x^*.

Consider, for example, for $N \gg 1$ the integral

$$I_0 = \int dx \, e^{-NF(x)} \, . \tag{E.7}$$

Using the saddle-point approximation this becomes

$$I_0 \approx \int dx \, \exp \left\{ -N \left[F(x^*) + \frac{1}{2} F''(x^*)(x - x)^2 \right] \right\}$$

$$= e^{-NF(x^*)} \cdot \sqrt{\frac{2\pi}{NF''(x^*)}} \tag{E.8}$$

where the "saddle point" x^* is determined by the condition that $e^{-NF(x)}$ has a maximum at x^*, i.e.

$$F'(x^*) = 0 \, . \tag{E.9}$$

If we apply this approximation to (E.6), we obtain instead of (E.9):

$$- [x_{2i+2} - f(x_{2i+1}^*)] f'(x_{2i+1}^*) + x_{2i+1}^* - f(x_{2i}) = 0 \tag{E.10a}$$

$$\rightarrow x_{2i+1}^* \approx f(x_{2i}) + [x_{2i+2} - f^2(x_{2i})] f'[f(x_{2i})] \tag{E.10b}$$

and

$$I \approx \exp\{[x_{2i+2} - f^2(x_{2i})]^2/2\bar{\sigma}^2\} \tag{E.11}$$

(where we have omitted all pre-exponential factors because they will cancel out in $\langle x_n \rangle$) and

$$\bar{\sigma}^2 = \sigma^2 + \{f'[f(x_{2i})]\}^2\sigma^2 . \tag{E.12}$$

Thus, $\bar{\sigma}$ depends on x_{2i} after one integration, i.e. when we repeat this procedure (see below) we will always encounter x dependent σ's and instead of (E.6), we should therefore consider from the very beginning

$$I = \int dx_{2i+1} \exp\{-[x_{2i+2} - f(x_{2i+1})]^2/2\sigma^2(x_{2i+1})$$

$$-[x_{2i+1} - f(x_{2i})]^2/2\sigma^2(x_{2i})\} . \tag{E.13}$$

In analogy to our previous calculation, we also obtain eq. (E.11) for this I, but with (E.12) replaced by

$$\bar{\sigma}^2(x_{2i}) = \sigma^2[f(x_{2i})] + \{f'[f(x_{2i})]\}^2 \cdot \sigma^2[x_{2i}] . \tag{E.14}$$

If we combine eqns. (E.5, 11, 14) and rescale and relabel the variables, i.e.

$$x_{2i} \equiv \bar{x}_i/a, \qquad (a = -|a|) \tag{E.15}$$

we obtain

$$\langle x_n \rangle \propto \int \prod_1^{n/2} d\bar{x}_i \bar{x}_{n/2} \prod_0^{n/2-1} P[\bar{x}_{i+1} - Tf(\bar{x}_i); \bar{\sigma}^2(\bar{\sigma}_i)] \tag{E.16}$$

where T is again the doubling operator

$$Tf(x) = af\left[f\left(\frac{x}{a}\right)\right] \tag{E.17}$$

and

$$\bar{\sigma}^2(x) = a^2\left\{\sigma^2\left[f\left(\frac{x}{a}\right)\right] + \left[f'\left[f\left(\frac{x}{a}\right)\right]\right]^2 \cdot \sigma^2\left(\frac{x}{a}\right)\right\} \equiv \hat{L}_f\sigma^2(x) \tag{E.18}$$

i.e. $\bar{\sigma}^2(x)$ is obtained by acting on $\sigma^2(x_n)$ with a linear operator $\hat{L}_f$. We note that the rescaling and relabeling were necessary to bring the expression (E.16) for $\langle x_n \rangle$ (after the odd variables had been integrated out) back into the *old form* (E.4) such that the whole renormalization-group transformation can be iterated.

After m renormalization steps we obtain finally

$$\langle x_n \rangle \propto \prod_1^{n/2^m} d\bar{x}_i \bar{x}_{n/2^m} \prod_0^{n/2^{m-1}} P[\bar{x}_{i+1} - T^m f(\bar{x}_i); \hat{L}_{T^{m-1}f} \dots \hat{L}_{Tf} \cdot \hat{L}_f \sigma^2 (\bar{x}_i)]. \tag{E.19}$$

For $m \gg 1$ we have again (see 3.53)

$$T^m f_R(x) = g(x) + r\delta^m a h(x) \quad \text{with} \quad r = R_\infty - R \tag{E.20}$$

and in analogy to (3.36–3.43)

$$\hat{L}_{T^{m-1}f} \cdot \dots \hat{L}_f \sigma^2(x) \cong \hat{L}_g^m \sigma^2(x) \cong \hat{\beta}^{2m} \hat{\sigma}^2(x) \tag{E.21}$$

where $\hat{\beta}^2$ and $\hat{\sigma}^2$ denote the largest eigenvalue and eigenfunction of $\hat{L}_g$, respectively. Thus, $\langle x_n \rangle$ can be written as

$$\langle x_n \rangle \propto \prod_1^{n/2^m} d\bar{x}_i \bar{x}_{n/2^m} \prod_0^{n/2-1} P[\bar{x}_{i+1} - g(\bar{x}_i) - r\delta^m a h(\bar{x}_i); \hat{\beta}^{2m} \hat{\sigma}^2(\bar{x}_i)]. \tag{E.22}$$

For the Liapunov exponent λ, this yields

$$\exp[n\lambda(r; \sigma)] = \left| \frac{d}{dx_0} \langle x_n \rangle \right| = \exp[(n/2^m) \lambda [r\delta^m; \sigma \cdot \hat{\beta}^m]] \tag{E.23}$$

where σ denotes the initial noise amplitude. If we set $\beta^m \cdot \sigma = 1$ and $\lambda(x; 1) = L(x)$, we obtain the desired scaling behavior for λ:

$$\lambda(r, \sigma) = \sigma^\theta L[r\sigma^{-\gamma}] \tag{E.24}$$

with

$$\theta = \log 2/\log \hat{\beta} = 0.367 \quad \text{and} \quad \gamma = \log \delta/\log \hat{\beta} = 0.815 . \tag{E.25}$$

Note that the numerical value for $\hat{\beta}$ ($\hat{\beta} = 6.618$) that was obtained as the solution of the eigenvalue equation

$$\hat{L}_g \hat{\sigma}^2(x) = \hat{\beta}^2 \hat{\sigma}^2(x) \tag{E.26}$$

agrees closely with the best value for μ ($\mu = 6.557$). This justifies our earlier treatment of external noise.

F   Shannon's Measure of Information

References: see Chapter 5

This short heuristic introduction into Shannon's measure of information should enable the reader to understand Chapters 2 and 5. For a more detailed treatment we recommend the book by Shannon and Weaver (1949).

Information Capacity of a Store

Fig. 157a) shows a system with two possible states. If the position of the points is unknown, a priori, and we learn that it is in the left box, say, we gain by definition information amounting to one bit. If we obtain this information, we save one question (with possible answer yes or no which we would have needed to locate the point). Thus, the maximum information content of a system with two states is one bit.

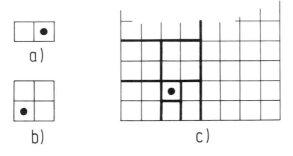

Fig. 157: Information capacity of a store: a) a box with two states. b) It takes two questions (and their answers) to locate a point in a system with four states: right or left? up or down? c) In order to locate a point on a checkerboard with $64 = 2^6$ states, one needs six questions.

For a box with four possible states, one needs two questions to locate the point, i. e. its maximum information content I is

$$I = 2 \text{ (bits)} \tag{F.1}$$

(we will drop the unit "bit" in the following).
This can be written as the logarithm to the base two (ld) of the number of possible states:

$$I = \text{ld } 4 . \tag{F.2}$$

According to Fig. 157 c, this logarithmic relation between the maximum information content I and the number of states N,

$$I = \text{ld } N \tag{F.3}$$

is true in general.

Information Gain

Let us now calculate the average gain of information if one learns the outcome of statistical events. Suppose we toss a coin such that heads or tails occur with equal probabilities

$$p_1 = p_2 = \frac{1}{2} . \tag{F.4}$$

The information I acquired by learning that the outcome of this experiment is heads, say, is

$$I = 1 \tag{F.5}$$

because there are two equally probable states as in Fig. 157 a. This result can be expressed via the $\{p_i\}$ as

$$I = - \left(\frac{1}{2} \text{ld } \frac{1}{2} + \frac{1}{2} \text{ld } \frac{1}{2} \right) \tag{F.6}$$

or

$$I = - \sum_i p_i \text{ld} p_i . \tag{F.7}$$

Eq. (F.7) can be generalized to situations where the p_i's are different:

$$p_1 \neq p_2 = 1 - p_1 . \tag{F.8}$$

It then gives the average gain of information if we toss a deformed coin many times.

Let $p_1 = r/q$, where r and q are mutually prime integers, and let us choose the number m of events such that mr/q is again an integer. The total number of distinct states which occur if one tosses a (deformed) coin m times is

$$N = \frac{m!}{(p_1 m)! \, (p_2 m)!} \tag{F.9}$$

where we eliminated, by division, the permutations that correspond to a rearrangement of equal events. The sequences hht and hht with $h =$ head and $t =$ tail, where the h's have been interchanged correspond to the same state. In the limit $m \rightarrow \infty$ we can use Stirling's formula, and, for the average information, gain eq. (F.3) yields

$$I = \frac{1}{m} \,\mathrm{ld}\, N = \frac{1}{m} \,\mathrm{ld}\, \left[\left(\frac{m}{e} \right)^m \left(\frac{e}{p_1 m} \right)^{p_1 m} \left(\frac{e}{p_2 m} \right)^{p_2 m} \right] =$$

$$= - \left(p_1 \,\mathrm{ld}\, p_1 + p_2 \,\mathrm{ld}\, p_2 \right) . \tag{F.10}$$

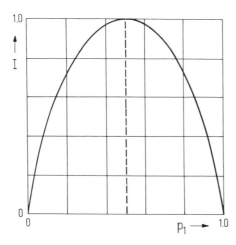

Fig. 158: $I\,(p)$ for an experiment with two possible outcomes. If $p_1 = 0$, we are sure that the outcome will be event 2, and we gain no information. The maximum information is acquired for $p_1 = p_2 = 1/2$, where the uncertainty of the outcome has its maximum and one learns most from the experiment.

This confirms eq. (F.7) and can be generalized to Shannon's result: If we, a priori, know only that $1 \ldots n$ events (or states of a system) occur with probabilities $\{p_i\}$ (such that $\sum\limits_{i=1}^{n} p_i = 1$) and we learn by a measurement that a certain event j has taken place (or the system occupies actually a certain state), then (if we repeat this measurement many times) we gain the average information

$$I = - \sum_{i=1}^{n} p_i \,\mathrm{ld}\, p_i . \tag{F.11}$$

G Period Doubling for the Conservative Hénon Map

References: see Chapter 6

Let us consider the quadratic area-preserving Hénon map

$$x_{n+1} = 1 - ax_n^2 - y_n \qquad \text{(G.1a)}$$

$$y_{n+1} = x_n \qquad \text{(G.1b)}$$

that describes (as we have seen in Chapter 1) a periodically kicked rotator for zero damping and small amplitudes. We want to show that this map (which represents a whole class of two-dimensional maps with a quadratic maximum) also leads to a cascade of period doublings, but with Feigenbaum constants that are larger than those for one-dimensional maps.

It is convenient to transform (G.1a, b) using

$$x_n = -\frac{2}{a}\bar{x}_n + \beta; \quad a\beta^2 + 2\beta - 1 = 0; \quad C = -a\beta \qquad \text{(G.2)}$$

into the form

$$\left.\begin{array}{l} y_{n+1} = x_n \\[2mm] x_{n+1} = 2Cx_n + 2x_n^2 - y_n \end{array}\right\} \equiv T\begin{pmatrix} x_n \\ y_n \end{pmatrix} \qquad \text{(G.3)}$$

(where we have omitted the bar notation).

We will first discuss the fixed points of T and T^2 and their stability, and finally introduce Helleman's renormalization scheme (Helleman, 1980), which sheds some light on the doubling mechanism and allows a convenient estimate of the relevant Feigenbaum constants.

The fixed points of T are

$$x_1^* = y_1^* = 0 \quad \text{and} \quad x_2^* = y_2^* = 1 - C \qquad \text{(G.4)}$$

and those of the second iterate T^2 where

$$T^2\begin{pmatrix} x_n \\ y_n \end{pmatrix} = \begin{cases} x_{n+2} = 2C[2Cx_n + 2x_n^2 - y_n] + 2[2Cx_n + 2x_n^2 - y_n]^2 - x_n \\[2mm] y_{n+2} = 2Cx_n + 2x_n^2 - y_n \end{cases} \qquad \text{(G.5)}$$

are the solution of

$$(Cx + x^2)^2 + C(Cx + x^2) - x = 0 . \tag{G.6}$$

To solve this equation it is noted that the fixed points (G.4) of T are also fixed points of T^2, i.e. (G.6) can be reduced to a quadratic equation with the solutions:

$$x^*_{3,4} = y^*_{3,4} = \frac{1}{2} \left[- (C + 1) \pm \sqrt{(C + 1)(C - 3)} \right] . \tag{G.7}$$

The stability of the fixed points is (by analogy to the one-dimensional case) determined by the eigenvalues $\lambda_{1,2}$ of the matrix of derivatives

$$L(x^*, y^*) = \begin{pmatrix} \dfrac{\partial T_x}{\partial x} & \dfrac{\partial T_x}{\partial y} \\[2mm] \dfrac{\partial T_y}{\partial x} & \dfrac{\partial T_y}{\partial y} \end{pmatrix}_{x^*, y^*} = \begin{pmatrix} 2C + 4x^* & -1 \\ 1 & 0 \end{pmatrix} \tag{G.8}$$

which are

$$\lambda_{1,2} = \frac{1}{2} \left[TrL \pm \sqrt{(TrL)^2 - 4} \right], \quad TrL = 2C + 4x^* . \tag{G.9}$$

Since T is an area-preserving map, $\det L = 1$, i.e. $\lambda_2 = 1/\lambda_1$. This leaves (apart from parabolic fixed points that we do not consider here because they are not generic) only two essentially different types of fixed points:

a) Hyperbolic fixed point: The λ's are real and $\lambda_1 > 1$ implies $\lambda_2 = 1/\lambda_1 < 1$, i.e. along the directions of the eigenvectors e_1, e_2 the behavior shown in Fig. 159 is found, which can be described by

$$T \begin{bmatrix} x \\ y \end{bmatrix} = \begin{bmatrix} x^* \\ y^* \end{bmatrix} + L \begin{bmatrix} \Delta x \\ \Delta y \end{bmatrix}$$

$$L \begin{bmatrix} \Delta x \\ \Delta y \end{bmatrix} = \begin{bmatrix} \lambda_1 & \Delta x \\ 1/\lambda_1 & \Delta y \end{bmatrix} \tag{G.10}$$

i.e. this fixed point is unstable since all points which are not on the stable manifold along e_2 are driven away from (x^*, y^*), and an infinite number of iterations is required to approach the fixed point along e_2:

$$\lim_{n \to \infty} L^n \begin{bmatrix} 0 \\ \Delta y \end{bmatrix} = \lim_{n \to \infty} \begin{bmatrix} 0 \\ (1/\lambda_1)^n \Delta y \end{bmatrix} = \begin{bmatrix} 0 \\ 0 \end{bmatrix} \tag{G.11}$$

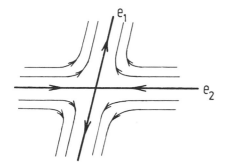

Fig. 159: Trajectories around of a hyperbolic fixed point with eigenvectors e_1 and e_2.

b) Elliptic fixed point: The λ's, as solutions of a quadratic equation, are complex conjugates and can be written as

$$\lambda_{1,2} = e^{\pm i\varphi} \quad \text{because} \quad \det L = \lambda_1^* \lambda_1 = 1 . \tag{G.12}$$

After an appropriate coordinate transformation, L can be written as a simple rotation:

$$L \begin{bmatrix} \Delta x \\ \Delta y \end{bmatrix} = \begin{bmatrix} \cos\varphi, & -\sin\varphi \\ \sin\varphi, & \cos\varphi \end{bmatrix} \begin{bmatrix} \Delta x \\ \Delta y \end{bmatrix} \tag{G.13}$$

and the fixed point is stable as shown in Fig. 160 because every point in its close vicinity remains there and is never driven away by applying L.

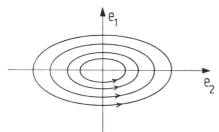

Fig. 160: Trajectories around of an elliptic fixed point.

Because, according to (G.9), the eigenvalues depend only on the trace of the linearized transformation matrix, we obtain the following criterion for stability:

$$|TrL| < 2 \rightarrow \text{stable}$$
$$\text{fixed point .} \tag{G.14}$$
$$|TrL| > 2 \rightarrow \text{unstable}$$

The stability of the fixed points (G.4) of T, therefore, becomes

$$x^* = y^* = 0 \quad \rightarrow |TrL| = |2C| \quad \rightarrow \text{stable for} \quad |C| < 1 \tag{G.15a}$$
$$x^* = y^* = 1-C \rightarrow |TrL| = 2|2-C| \rightarrow \text{unstable for} \; |C| > 1 \tag{G.15b}$$

For T^2 we have by analogy to the chain rule $df^2(x)/dx = f'[f(x)]f'(x)$ in the one-dimensional case:

$$Tr\, L_{T^2} = Tr\,[L_T(x_3^*\, y_3^*) \cdot L_T(x_4^*, y_4^*)] \tag{G.16}$$

$$= 2\,[-\,2\,(C+1)(C+3)+1] = \begin{cases} +1 & \text{for} \quad C = -1 \\ -1 & \text{for} \quad C = 1 - \sqrt{5} \end{cases}$$

where we denoted the functional matrix of T^2 by L_{T^2} and used $(x_3^*, y_3^*) = T(x_4^*, y_4^*)$.

Collecting (G.15–16) together, we find: (x_1^*, y_1^*) is an attractor of period 1 and is stable for $-1 < C < 1$, and $(x_{3,4}^*, y_{3,4}^*)$ is an attractor of period 2 and stable for $1 - \sqrt{5} < C < -1$. We, therefore, see the beginning of a bifurcation cascade.

Let us now demonstrate the self-similarity which leads to the whole sequence of period doublings by introducing Hellemann's renormalization scheme.

Its starts from (G.3) which can be written as

$$x_{n+1} + x_{n-1} = 2C x_n + 2 x_n^2 \,. \tag{G.17}$$

A linearization of this equation around the fixed points of period two,

$$x_n^* = \frac{1}{2}\,[-(C+1) + (-1)^n \sqrt{(C+1)(C-3)}\,]\,; \quad n = 0,1,2,3 \tag{G.18}$$

yields

$$\Delta x_{n+1} + \Delta x_{n-1} = (2C + 4x_n^*)\,\Delta x_n - 2\,(\Delta x_n)^2 \,. \tag{G.19}$$

If we add (G.19), then for $n = 2m+1$ and $n = 2m-1$ we obtain

$$\Delta x_{2m+2} + \Delta x_{2m-2} = -2\Delta x_{2m} + (2C + 4x_0^*)\,[\Delta x_{2m+1} + \Delta x_{2m-1}$$
$$+ 2\,[(\Delta x_{2m-1})^2 + (\Delta x_{2m+1})^2]\,. \tag{G.20}$$

Now we take (G.19) at $n = 2m$,

$$\Delta x_{2m+1} + x_{2m+1} = (2C + 4x_1^*)\,\Delta x_{2m} + 2\,(\Delta x_{2m})^2 \tag{G.21}$$

and add it to (G.20):

$$\Delta x_{2m+2} + \Delta x_{2m-2} = 2C'\Delta x_{2m} + 2a\,(\Delta x_{2m})^2 + O\,[(\Delta x)^3]\,. \tag{G.22}$$

This equation can be put into the same as (G.17) by rescaling $x_m' \equiv a\,\Delta x_{2m}$:

$$x_{m+1}' + x_{m-1}' = 2C' x_m' + 2 x_m'^2 \tag{G.23}$$

where

$$C' = 2(C + 2x_1^*)(C + 2x_0^*) - 1 = 2C^2 + 4C + 7 \tag{G.24}$$

$$a = 2(C + 2x_1^*) + 2(C + 2x_0^*)^2 \ . \tag{G.25}$$

The meaning of eq. (G.23) is as follows: If the two-dimensional map is developed to second order around the two-cycle and the result is rescaled, one obtains the old map, i. e. the stability of $x^* = y^* = 0$ for $|C| < 1$ implies (because of the similarity of (G.17) and (G.23)) the stability of $x^{*\prime} = y^{*\prime} = \Delta x = \Delta y = 0$ i. e. of the two-cycle for $|C| = | -2C^2 + 4C + 7| < 1$ or $1 - \sqrt{5} < C < -1$.

Repeating this argument we see that (G.23) also holds for the derivatives around a four-cycle, etc. A cascade of bifurcations with cycles of period 2^n is obtained which are stable for $C_{n-1} < C < C_n$ where

$$C_{n-1} = 2C_n^2 + 4C_n + 7 \ . \tag{G.26}$$

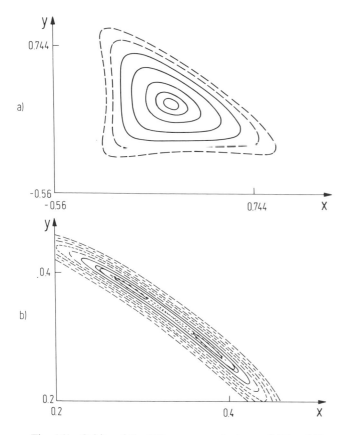

Fig. 161: Orbits of the Hénon map: a) at $a = 0.95$ and b) at $a = 3.02$ (after Bountis, 1981).

The bifurcation points accumulate at C_∞ which is determined by

$$C_\infty = 2C_\infty + 4C_\infty + 7 \rightarrow C_\infty = -1.2656(1.266311\ldots) \qquad \text{(G.27)}$$

which yields

$$a = a(C_\infty) = -4.128(4.018077\ldots) \qquad \text{(G.28)}$$

and the Feigenbaum constant δ

$$C_n = C_\infty + A\delta^{-n} \quad \text{with} \quad \delta = 9.06(8.72109\ldots) \qquad \text{(G.29)}$$

where the numbers in parenthesis give the best current numerical values for the constants.

Fig. 161 shows the orbits of the Hénon map (G.1 a, b) near a stable fixed point and after the first bifurcation.

Remarks and References

Since the introductory review articles and books mentioned below already contain extensive lists of original literature (there are already 300 references in the articles by Hellemann), no attempt has been made to cover all references. Authors, whose work has inadvertently not been cited, are asked for their forgiveness. The references and further reading material are listed according to the individual chapters.

Introduction

Original articles and general sources:

Abraham, R. H., and Shaw, C. D. (1981): *Dynamics — The Geometry of Behaviour.* Aeriel Press, Santa Cruz.

Arnold, V. I. (1963), "Small Denominators II, Proof of a Theorem of A. N. Kolmogorov on the Preservation of Conditionally-Periodic Motions Under a Small Perturbation of the Hamiltonian", *Russ. Math. Surveys* **18**, 5.

Arnold, V. I., and Avez, A. (1968): *Ergodic Problems of Classical Mechanics.* Benjamin, New York.

Balescu, R. (1975): *Equilibrium and Nonequilibrium Statistical Mechanics.* Wiley, New York.

Bergé, P., Pomeaú, Y., and Vidal, Ch. (1984): *L'Ordre dans le Chaos.* Hermann, Paris.

Campbell, D., and Rose, H. (eds.) (1983). "Proc. of the International Conference on 'Order in Chaos' in Los Alamos", *Physica* **7D**, 1.

Chandrasekhar, S. (1961): *Hydrodynamic and Hydromagnetic Stability.* Clarendon Press, Oxford.

Chirikov, B. V. (1980): "A Universal Instability of Many Dimensonal Oscillator Systems", *Phys. Rep.* **52**, 463.

Collet, P., and Eckmann, J. P. (1980): *Iterated Maps of the Interval al Dynamical Systems.* Birkhäuser, Boston.

Coullet, P., and Tresser, J. (1978), "Iterations d'Endomorphismes et Groupe de Renormalisation", *C. R. Hebd. Seances Acad. Sci., Ser. A* **287**, 577, and *J. Phys. (Paris) Coll.* **39**, C 5–25 (1978).

Cvitanovich, P. (ed.) (1984): *Universality in Chaos.* A reprint selection, Adam Hilger, Bristol.

Eckmann, J. P. (1981), "Roads to Turbulence in Dissipative Dynamical Systems", *Rev. Mod. Phys.* **53**, 643.

Feigenbaum, M. J. (1978), "Quantitative Universality for a Class of Nonlinear Transformations", *J. Stat. Phys.* **19**, 25.

Feigenbaum, M. (1980): *Universal Behaviour in Nonlinear Systems.* Los Alamos Science.

Garrido, L. (ed.) (1982): *Dynamical Systems and Chaos.* Lect. Notes in Physics **179**, Springer, Berlin-Heidelberg-New York-Tokyo.

Gorel, D. G. and Roessler, O. E. (eds.) (1978), "Bifurcation Theory in Scientific Disciplines", *Ann. N. Y. Acad. Sci.* **306**.

Grossmann, S., and Thomae, S. (1977), "Invariant Distributions and Stationary Correlation Functions of One-Dimensional Discrete Processes", *Z. Naturforsch.* **32 A**, 1353.

Guckenheimer, J., and Holmes, P, (1983): *Nonlinear Oscillations, Dynamical Systems and Bifurcations of Vector Fields.* Springer, Berlin-Heidelberg-New York-Tokyo.

Haken, H. (ed.) (1981): *Chaos and Order in Nature.* Springer, Berlin-Heidelberg-New York-Tokyo.

Haken, H. (1982): *Synergetics.* 2nd Ed. Springer, Berlin-Heidelberg-New York-Tokyo.

Haken, H. (ed.) (1982): *Evolution of Order and Chaos.* Springer, Berlin-Heidelberg-New York-Tokyo.

Haken, H. (1984): *Advanced Synergetics.* Springer, Berlin-Heidelberg-New York-Tokyo.

Hao Bai-Lin (1984): *Chaos.* World Scientific, Singapore.

Hellemann, R. H. G. (1980), in E. G. D. Cohen (ed.): *Fundamental Problems in Statistical Mechanics.* Vol. V. North-Holland, Amsterdam-New York, p. 165.

Hellemann, R. H. G. (ed.) (1980), "Proc. of the 1979 Int. Conf. on Nonlinear Dynamics", *Ann. N. Y. Acad. Sci.* **603**.

Hellemann, R. H. G., and Ioos, G. (eds.) (1983): *Les Houches Summerschool on "Chaotic Behaviour in Deterministic Systems".* North-Holland, Amsterdam.

Hu, B. (1982), "Introduction to Real Space Renormalization Group Methods in Critical and Chaotic Phenomena", *Phys. Rep.* **31**, 233.

Kolmogorov, A. N. (1954), "On Conservation of Conditionally-Periodic Motions for a Small Change in Hamilton's Function", *Dokl. Akad. Nauk. USSR* **98**, 525.

Landau, L. D. (1944), *C. R. Dokl. Acad. Sci. USSR* **44**, 311.

Landau, L. D., and Lifshitz, E. M. (1959): *Fluid Mechanics,* Pergamon, Oxford.

Lichtenberg, A. J., and Liebermann, M. A. (1983): *Regular and Stochastic Motion.* Springer, Berlin-Heidelberg-New York-Tokyo.

Lorenz, E. N. (1963), "Deterministic Nonperiodic Flow", *J. Atmos. Sci.* **20**, 130.

Manneville, P., and Pomeau, Y. (1979), "Intermittency and the Lorenz Model", *Phys. Lett.* **75 A.** 1.

May, R. M. (1976), "Simple Mathematical Models with very Complicated Dynamics", *Nature* **261**, 459.

Monin, A. S. and Yaglom, A. N. (1978): *Statistical Fluid Mechanics.* MIT Press, Cambridge. Mass.

Moser, J. (1967), "Convergent Series Expansions of Quasi-Periodic Motions", *Math. Ann.* **169**, 163.

Moser, J. (1973): *Stable and Random Motion in Dynamical Systems.* Princeton Univ. Press.

Moser, J. (1978), "Is the Solar System Stable?", *Math. Intelligencer* **1**, 65.

Newhouse, S., Ruelle, D., and Takens, F. (1978), "Occurrence of Strange Axiom-A Attractors near Quasiperiodic Flow on T^m, m $\geq$ 3", *Commun. Math. Phys.* **64**, 35.

Ott, E. (1981), "Strange Attractors and Chaotic Motions of Dynamical Systems", *Rev. Mod. Phys.* **53**, 655.

Peitgen, H. O., and Richter, P. H. (1986): *The Beauty of Fractals.* Springer, Berlin-Heidelberg-New York-Tokyo.

Poincaré, H. (1892): *Les Méthodes Nouvelles de la Méchanique Celeste,* Gauthier-Villars, Paris; in English (1967): *N.A.S.A Translation TT F-450/452.* U.S. Fed. Clearinghouse; Springfield, VA, USA.

Ruelle, D., and Takens, F. (1971), "On the Nature of Turbulence", *Commun. Math. Phys.* **20**, 167.

Ruelle, D. (1980), "Strange Attractors", *Math. Intelligencer* **2**, 126.

Shaw, R. S. (1981), "Strange Attractors, Chaotic Behaviour and Information Flow", *Z. Naturforsch.* **36 A**, 80.

Swinney, H. L. and Gollub, J. P. (1981): *Hydrodynamic Instabilities and the Transition to Turbulence.* Springer, Berlin-Heidelberg-New York-Tokyo.

Vidal, C., and Pacault, A. (1981): *Nonlinear Phenomena in Chemical Dynamics.* Springer, Berlin-Heidelberg-New York-Tokyo.

Zaslavsky, G. M. (1981), "Stochasticity in Quantum Systems", *Phys. Rep.* **80**, 157.

More information about the systems quoted in Table 1 can be found in the following references:

[1]   D. D'Humières, M. R., Beasly, B. A., Humberman and A. Libchaber, "Chaotic States and Routes to Chaos in the Forced Pendulum", *Phys. Rev.* **26 A** (1982) 3483.

[2]   A. Libchaber and J. Maurer in T. Riste (ed.): *Nonlinear Phenomena at Phase Transitions and Instabilities.* NATO Adv. Study Inst., Plenum Press, New York 1982; see also the book edited by Swinney and Gollub, mentioned under general sources.

[3]   H. Haken, "Analogy between Higher Instabilities in Fluids and Lasers", *Phys. Lett.* **53 A** (1975) 77.

[4]   F. A. Hopf, D. L. Kaplan, H. M. Gibbs and R. L. Shoemaker, "Bifurcations to Chaos in Optical Bistability", *Phys. Rev.* **25 A** (1982) 2172.

[5]   M. Cirillo and N. F. Pedersen, "On Bifurcations and Transitions to Chaos in a Josephson Junction", *Phys. Lett.* **90 A** (1982) 150.

[6]   R. H. Simoyi, A. Wolf and H. L. Swinney: "One Dimensional Dynamics in a Multicomponent Chemical Reaction", *Phys. Rev. Lett.* **49** (1982) 245.

[7–8] See the review article by Hellemann, mentioned under general sources.

[9]   J. M. Wersinger, J. M. Finn and E. Ott in G. Laval and D. Gresillon (eds.): *Intrinsic Stochasticity in Plasmas.* Les Editions de Physique, Courtaboeuf, Orsay, France 1980.

[10] See the article by May, mentioned under general sources.

[11] L. Glass, M. R. Guevara and A. Shrier, "Bifurcation and Chaos in a Periodically Stimulated Cardiac Oscillator", *Physica* **7D** (1983) 89, and A. T. Winfree, *Sci. Am.* **248**, No. 5 (1983).

1   Experiments and Simple Models

More information on the systems presented in Table 2 can be found in the following references:

Dubois, M., and Bergé, P. (1981). "Instabilities de Couche Limite dans un Fluide en Convection. Evolution vers la Turbulence", *J. Phys. (Paris)* **42**, 167.

Epstein, I. R., Kustin, K., de Kepper, P., and Orban, M. (1983), "Oscillating Chemical Reactions", *Sci. Am.* **248**, No. 3.

Haken, H. (1975), "Analogy between Higher Instabilities in Fluids and Lasers", *Phys. Lett.* **53 A,** 77.

Haken, H. (1982): *Synergetics.* 2nd Edition. Springer, Berlin-Heidelberg-New York-Tokyo.

Haken, H. (1984): *Advanced Synergetics.* Springer, Berlin-Heidelberg-New York-Tokyo.

Hénon, M., and Heiles, C. (1964), "The Applicability of the Third Integral of the Motion: Some Numerical Results", *Astron, J.* **69,** 73.

D'Humieres, D., Beasly, M. R., Huberman, B. A., and Libchaber, A. (1982), "Chaotic States and Routes to Chaos in the Forced Pendulum", *Phys. Rev.* **A 26,** 3483.

Ma, S. (1976): *Modern Theory of Critical Phenomena.* Benjamin, New York.

McLaughlin, J. B., and Martin, P. C. (1975), "Transition to Turbulence of a Statically Stressed Fluid System", *Phys. Rev.* **A 12,** 186.

Libchaber, A., and Maurer, J. (1982) in Riste, T. (ed.): *Nonlinear Phenomena at Phase Transitions and Instabilities.* NATO Advanced Study Inst., Plenum Press, New York.

Lorenz, E. N. (1963), "Deterministic Nonperiodic Flow", *J. Atmos. Sci.* **20,** 130.

Roux, J. C., Rossi, A., Bachelart, S., and Vidal, C. (1981): "Experimental Observations of Complex Behaviour During a Chemical Reaction", *Physica* **2D,** 395.

Saltzman, B. (1981), "Finite Amplitude Free Convection as an Initial Value Problem I", *J. Atmos. Sci.* **19,** 329.

Simoyi, R. H., Wolf, A., and Swinney, H. L. (1982), "One-Dimensional Dynamics in a Multicomponent Chemical Reaction", *Phys. Rev. Lett.* **49,** 245.

Swinney, H. L., and Gollub, J. P. (1978), "The Transition to Turbulence", *Phys. Today* **31 (8),** 41.

Swinney, H. L., and Gollub, J. P. (eds.) (1980): *Hydrodynamic Instabilities and the Transition to Turbulence.* Springer, Berlin-Heidelberg-New York-Tokyo.

Tyson, J. J. (1976): *The Belousov-Zhabotinsky Reaction.* Lecture Notes in Biomathematics **22,** Springer, Berlin-Heidelberg-New York-Tokyo.

Vidal, C., Roux, J. C., Bachelart, S., and Rossi, A., in Hellemann, R. H. G. (ed.) (1980): "Proc. of the 1979 Int. Conf. on Nonlinear Dynamics", *Ann. N.Y. Acad. Sci.* **603.**

Wegmann, K., and Roessler, O. E., (1978). "Different Kinds of Chaos in the Belousov-Zhabotinsky Reaction", *Z. Naturforsch.* **33 a.** 1179.

Wilson, K. G., and Kogut, J. (1974), "The Renormalization Group and the ε-Expansion", *Phys. Rep.* **C 12,** 75.

The periodically kicked rotator and related topics are studied in the following articles:

Campbell, D., and Rose, H., (eds.) (1983), "Proc. of the International Conference on "Order in Chaos" in Los Alamos", *Physica* **7 D,** 1.

Chandrasekhar, S. (1961): *Hydrodynamic and Hydromagnetic Stability,* Clarendon Press, Oxford.

Chirikov, B. V. (1979), "A Universal Instability of Many Dimensional Oscillator Systems", *Phys. Rep.* **52,** 463.

Hénon, M. (1976), "A Two Dimensional Map with a Strange Attractor". *Commun. Math. Phys.* **50,** 69.

Ott, F. (1981), "Strange Attractors and Chaotic Motions of Dynamical Systems", *Rev. Mod. Phys.* **53,** 655.

Zaslavsky, G. M. (1978), "The Simplest Case of a Strange Attractor", *Phys. Lett.* **69 A,** 145.

Zaslavski, G. M., and Rachki, Kh. R. Ya. (1979), "Singularities of the Transition to Turbulent Motion", *Sov. Phys. J.E.T.P.* **49,** 1039.

2   Piecewise Linear Maps and Deterministic Chaos

The Bernoulli shift is discussed in:

Billingsley, O. (1964): *Ergodic Theory and Information.* Wiley, New York.

Shaw, R. S. (1981), in Haken, H. (ed.): *Chaos and Order in Nature.* Springer, Berlin-Heidelberg-New York-Tokyo.

A proof that almost all irrational numbers in [0,1] in their binary representation contain any finite sequence infinitely many times can be found in:

Hardy, G. H., and Wright, E. M. (1938): *The Theory of Numbers,* Oxford University Press, Oxford, p. 125.

More on the behavior of piecewise linear maps can be found in:

Eckmann, J. P., and Ruelle, D. (1985), "Ergodic Theory of Chaos and Strange Attractors", *Rev. Mod. Phys.* **57**, 617.

Grossmann, S., and Thomae, S. (1977), "Invariant Distributions and Stationary Correlation Functions of One-Dimensional Discrete Processes", *Z. Naturforsch.* **32 A,** 1353.

Shaw, R. S. (1981), "Strange Attractors, Chaotic Behaviour and Information Flow", *Z. Naturforsch.* **36 A**, 80.

Deterministic diffusion has been studied for example in the following papers:

Geisel, T., and Nierwetberg, J. (1982). "Onset of Diffusion and Universal Scaling in Chaotic Systems", *Phys. Rev. Lett.* **48**, 7.

Grossmann, S., and Fujisaka, H. (1982), "Diffusion in Discrete Nonlinear Systems", *Phys. Rev.* **26 A**, 1779.

Grossmann, S. (1982) in Haken, H. (ed.): *Evolution of Order and Chaos.* Springer, Berlin-Heidelberg-New York-Tokyo.

Haken, H. (1982): *Synergetics.* 2nd Ed. Springer, Berlin-Heidelberg-New York-Tokyo.

3   Universal Behavior of Quadratic Maps

Original articles and general sources:

Collet, P., and Eckmann, J. P. (1980): *Iterated Maps of the Interval as Dynamical Systems,* Birkhäuser, Boston.

Collet, P., Eckmann, J. P., and Lanford, O. E. (1980), "Universal Properties of Maps on an Interval", *Commun. Math. Phys.* **76**, 211.

Coullet, P., and Tresser, J. (1978), "Iterations d'Endomorphismes et Groupe de Renormalisation", *C. R. Hebd, Seances Acad. Sci. Series A* **287,** 577, and *J. Phys. (Paris)* **C5**, 25 (1978).

Derrida, B., Gervois, A., and Pomeau, Y. (1979), "Universal metric Properties of Bifurcations and Endomorphisms", *J. Phys.* **12 A**, 269.

Feigenbaum, M. J. (1978), "Quantitative Universality for a Class of Nonlinear Transformations", *J. Stat. Phys.* **19**, 25.

Feigenbaum, M. J. (1979), "The Universal Properties of Nonlinear Transformations", *J. Stat. Phys.* **21**, 669.

Grossmann, S., and Thomae, S. (1977), "Invariant Distributions and Stationary Correlation Functions of One-Dimensional Discrete Processes", *Z. Naturforsch.* **32 A,** 1353.

Hubermann, B. A., and Rudnick, J. (1980), "Scaling Behaviour of Chaotic Flows", *Phys. Rev. Lett.* **45,** 154.

May, R. M. (1976), "Simple Mathematical Models with very Complicated Dynamics", *Nature* **261,** 459.

Metropolis, M., Stein, M. L., and Stein, P. R. (1973), "On Finite Limit Sets for Transformations of the Unit Interval", *J. Combinatorial Theory (A)* **15,** 25.

Lanford, O. E., (1982): "A Computer Assisted Proof of the Feigenbaum Conjectures", *Bull. Am. Math. Soc.* **6,** 427.

Peitgen, H. O., and Richter, P. H. (1984): *Harmonie in Chaos und Kosmos,* and *Morphologie komplexer Grenzen; Bilder aus der Theorie dynamischer Systeme.*
Both catalogues can be obtained from: Forschungsschwerpunkt: Dynamische Systeme, Universität Bremen, D-2800 Bremen, F.R.G.

The Hausdorff dimension and related topics are discussed in:

Grassberger, P. (1981), "On the Hausdorff Dimension of Fractal Attractors", *J. Stat. Phys.* **19,** 25.

Mandelbrot, B. B. (1982): *The Fractal Geometry of Nature,* Freeman, San Francisco.

The universal properties of the power spectrum are calculated in:

Collet, P., and Eckmann, J. P. (1980). *Iterated Maps of Interval as Dynamical Systems.* Birkhäuser, Boston.

Feigenbaum, M. J. (1979), "The Onset Spectrum of Turbulence", *Phys. Lett.* **74 A,** 375.

Feigenbaum, M. J. (1980), "The Transition to Aperiodic Behaviour in Turbulent Systems", *Commun. Math. Phys.* **77,** 65.

Nauenberg, M., and Rudnick, J. (1981), "Universality and the Power Spectrum at the Onset of Chaos", *Phys. Rev.* **B 24,** 439.

The influence of external noise on period doubling is, e.g., discussed in:

Crutchfield, J. P., Nauenberg, M., and Rudnick, J. (1981). "Scaling for External Noise at the Onset of Chaos", *Phys. Rev. Lett.* **46,** 933.

Crutchfield, J. P., Farmer, J. D., and Hubermann, B. A. (1982), "Fluctuations and Simple Chaotic Dynamics", *Phys. Rep.* **92,** 45.

Feigenbaum, M. J., and Hasslacher, B. (1982), "Irrational Decimations and Path Integrals for External Noise", *Phys. Rev. Lett.* **49,** 605.

Shraiman, B., Wayne, C. E., and Martin, P. C. (1981), "Scaling Theory for Noisy Period-Doubling Transition to Chaos", *Phys. Rev. Lett.* **46,** 935.

The behavior of the logistic map for $r > r_\infty$ is investigated in:

Derrida, B., Gervois, A., and Pomeau, Y. (1978), "Iteration of Endomorphisms on the Real Axis and Representation of Numbers", *Ann. Inst. Henri Poincaré* **29 A**, 305.

Farmer, J. D. (1985), "Sensitive Dependence on Parameters in Nonlinear Dynamics", *Phys. Rev. Lett.* **55**, 351.

Farmer, J. D. (1986), in G. Mayer Kress (ed.): *Dimensions and Entropies in Chaotic Systems*, p. 54. Springer. Berlin-Heidelberg-New York-Tokyo.

Feigenbaum, M. J. (1978), "Quantitative Universality for a Class of Nonlinear Transformations", *J. Stat. Phys.* **19**, 25.

Geisel, T., and Nierwetberg, J. (1981), "Universal Fine Structure of the Chaotic Region in Period-Doubling Systems", *Phys. Rev. Lett.* **47**, 975.

Guckenheimer, J. (1980), "One Dimensional Dynamics", *Ann. N.Y. Acad. Sci.* **357**, 343.

Jacobson, M. V. (1981), "Absolutely Continuous Invariant Measure for One-Parameter Families of One-Dimensional Maps", *Comm. Math. Phys.* **81**, 39.

Li, T., and Yorke, J. A. (1975), "Period Three implies Chaos", *Ann. Math. Monthly* **82**, 985.

Metropolis, N., Stein, M. L., and Stein, P. R. (1973), "On Finite Limit Sets for Transformations on the Unit Interval", *Jour. of Combinatorial Theory* **15**, 25.

Sarkovskii, A. N. (1964), "Coexistence of Cycles of a Continuous Map of a Line into Itself", *Ukr. Mat. Z.* **16**, 61.

Thomae, S., and Grossmann, S. (1981), "Correlations and Spectra of Periodic Chaos Generated by the Logistic Parabola", *J. Stat. Phys.* **26**, 485.

More on Sarkovskii's theorem and the Schwarzian derivative can be found in the book by Collet and Eckmann mentioned in the list of general references of the introduction and in:

Lanford, O. E. (1980), "Smooth Transformations of Intervals", *Seminare BOURBAKI* No. 563.

The references for the experiments quoted in the text are:

Lauterborn, W., and Cramer, E. (1981), "Subharmonic Route to Chaos Observed in Acoustics", *Phys. Rev. Lett.* **47**, 1145.

Lauterborn, W. (1982), "Cavitation Bubble Dynamics. New Tools for an Intricate Problem", *Appl. Sci. Research* **38**, 165.

Lauterborn, W., and Mayer-Ilse, W. (1986), "Chaos", *Physik in unserer Zeit* **17**, 177.

Lauterborn, W., and Suchla, E. (1984), "Bifurcation Superstructure in a Model of Acoustical Turbulence", *Phys. Rev. Lett.* **53**, 2304.

Libchaber, A., and Maurer, J. (1980), "Une Experience de Rayleigh-Bénard de Géometrie Reduite; Multiplication, Accrochage et Démultiplication de Fréquences", *J. Phys. (Paris) Coll.* **41**, C 3–51.

Libchaber, A., and Maurer J. (1982) in Riste, T. (ed.): *Nonlinear Phenomena at Phase Transitions and Instabilities.* NATO Adv. Study Institute, Pergamon Press, New York.

Linsay, P. S. (1981), "Period Doubling and Chaotic Behaviour in a Driven Anharmonic Oscillator", *Phys. Rev. Lett.* **47**, 1349.

Rollins, R. W., and Hunt, E. R. (1982), "Exactly Solvable Model of a System Exhibiting Universal Chaotic Behaviour", *Phys. Rev. Lett.* **49**, 1295.

Testa, J., Pérez, J., and Jeffries, C. (1982), "Evidence for Universal Chaotic Behaviour of a Driven Nonlinear Oscillator". *Phys. Rev. Lett.* **48**, 714.

A renormalization-group treatment of second-order phase transitions can, e.g., be found in:

Ma, S. K. (1976): *Modern Theory of Critical Phenomena.* Benjamin, New York.

The Feigenbaum route has been observed in many experiments. Some more references are:

Arecchi, F. T., Meucci, R., Puccioni, G., and Tredice, J. (1982), "Experimental Evidence for Subharmonic Bifurcations, Multistability and Turbulence in a Q-switched Gas Laser", *Phys. Rev. Lett.* **49**, 1217.

Brun, E., Derighetti, D., Holzner, R., and Meier, D. (1983), "The NMR-Laser — a Nonlinear Solid State System Showing Chaos", *Helv. Physica Acta* **56**, 852.

Cirillo, M., and Pedersen, N. F. (1982), "On Bifurcations and Transition to Chaos in a Josephson Junction", *Phys. Lett.* **90 A**, 150.

Giglio, M., Musazzi, S., and Perine, V. (1981): "Transition to Chaotic Behaviour via a Reproducible Sequence of Period-Doubling Bifurcations", *Phys. Rev. Lett.* **47**, 243.

Gollub, J. P., and Swinney, H. L. (1975), "Onset of Turbulence in a Rotating Fluid", *Phys. Rev. Lett.* **35**, 927.

Gollub, J. P., and Benson, S. V. (1980), "Many Routes to Turbulent Convection", *J. Fluid Mech.* **100**, 499.

Hopf, F. A., Kaplan, D. L., Gibbs, H. M., and Shoemaker, R. L. (1982), "Bifurcations to Chaos in Optical Bistability", *Phys. Rev.* **25 A**, 2171.

Klinker, T., Meyer-Ilse, W., and Lauterborn, W. (1984), "Period Doubling and Chaotic Behaviour in a Driven Toda Oscillator", *Phys. Lett.* **101 A**, 371.

Libchaber, A., Laroche, C., and Fauve, S. (1982), "Period Doubling in Mercury, a Qualitative Measurement", *J. Phys. (Paris) Lett.* **43**, L211.

Pfister, G. (1984), "Period Doubling in Rotational Taylor-Couette Flow", *Proc. 2nd Int. Symp. on Appl. of Laser Anemometry to Fluid Mechanics.* Lisboa.

Simoyi, R. H., Wolf, A., and Swinney, H. L. (1982), "One-Dimensional Dynamics in a Multicomponent Chemical Reaction", *Phys. Rev. Lett.* **49**, 245.

Smith, C. W., Tejwani, M. J., and Farris, D. A. (1982), "Bifurcation Universality for First Sound Subharmonic Generation in Superfluid Helium-4", *Phys. Rev. Lett.* **48**, 429.

4   The Intermittency Route to Chaos

The intermittency route has been investigated in the following pioneering articles:

Manneville, P., and Pomeau, Y. (1979), "Intermittency and the Lorenz Model", *Phys. Lett.* **75 A**, 1.

Manneville, P., and Pomeau, Y. (1980), "Different Ways to Turbulence in Dissipative Dynamical Systems", *Physica* **1 D**, 219.

Pomeau, Y., and Manneville P. (1979), in G. Laval, and D. Gresillon (eds.): *Intrinsic Stochasticity in Plasmas.* Ed. de Physique, Orsay, p. 239.

Pomeau, Y., and Manneville, P. (1980), "Intermittent Transition to Turbulence in Dissipative Dynamical Systems", Comm. Math. Phys. **74, 189.**

Influence of external noise on intermittency:

Eckmann, J. P., Thomas, L., and Wittwer, P. (1981), "Intermittency in the Presence of Noise", *J. Phys.* **14 A**, 3153.

Hirsch, J. E., Hubermann, B. A., and Scalapino, D. J. (1982), "Theory of Intermittency", *Phys. Rev.* **25 A**, 519.

Hirsch, J. E., Nauenberg, M., and Scalapino, D. J. (1982), "Intermittency in the Presence of Noise: A Renormalization Group Formulation", *Phys. Lett.* **87 A**, 391.

Hu, B., and Rudnick, J. (1982), "Exact Solutions of the Feigenbaum Renormalization Group Equations for Intermittency", *Phys. Rev. Lett.* **48**, 1645.

More on $1/f$-noise can be found in:

Ben-Mizrachi, A., Procaccia, I., Rosenberg, N., Schmidt, A., and Schuster, H. G. (1985). "Real and Apparent Divergencies in Low-Frequency Spectra of Nonlinear Dynamical Systems", *Phys. Rev.* **A 31**, 1830.

Dutta, P., and Horn, P. M. (1981), "Low-Frequency Fluctuations in Solids", *Rev. Mod. Phys.* **53**, 497.

Manneville, P. (1980), "Intermittency, Self-Similarity ad $1/f$-Spectrum in Dissipative Dynamical Systems", *J. Phys. (Paris)*, **41**, 1235.

Procaccia, I., and Schuster, H. G. (1983), "Functional Renormalization Group Theory of Universal $1/f$-Noise in Dynamical Systems". *Phys. Rev.* **28 A**, 1210.

Wolf, D. (1978): *Noise in Physical Systems. Series in Electrophysics* **2**. Springer, Heidelberg-New York.

Experiments in which intermittency has been observed:

Bergé, P., Dubois, M., Manneville, P., and Pomeau, Y. (1980), "Intermittency in Rayleigh-Bénard Convection", *J. Phys. (Paris) Lett.* **41**, L-344.
Dubois, M., Rubio, M. A., and Bergé, P. (1983), "Experimental Evidence of Intermittencies Associated with a Subharmonic Bifurcation", *Phys. Rev. Lett.* **51**, 1446.
Jeffries, C., and Pérez, J. (1982), "Observation of a Pomeau-Manneville Intermittent Route to Chaos in a Nonlinear Oscillator", *Phys. Rev.* **26 A**, 2117.
Pomeau, Y., Roux, J. C., Rossi, A., Bachelart, S., and Vidal, C. (1981), "Intermittent Behaviour in the Belousov-Zhabotinsky Reaction", *J. Phys. (Paris) Lett.* **42**, L-271.
Yeh, W. J. and Kao, Y. H. (1983), "Intermittency in Josephson Junctions", *Appl. Phys. Lett.* **42**, 299.

5   Strange Attractors in Dissipative Dynamical Systems

Introduction and definition of strange attractors:

Eckmann, J. P. (1981), "Roads to Turbulence in Dissipative Dynamical Systems", *Rev. Mod. Phys.* **53**, 643.
Eckmann, J. P., and Ruelle, D. (1985), "Ergodic Theory of Chaos and Strange Attractors", *Rev. Mod. Phys.* **57**, 617.
Hénon, M. (1976), "A Two-Dimensional Map with a Strange Attractor", *Commun. Math. Phys.* **50**, 69.
Lanford, O. E. (1977), "Turbulence Seminar", in P. Bernard and T. Rativ (eds.): *Lecture Notes in Mathematics* **615**. Springer, Heidelberg-New York, p. 114.
Lorenz, E. N. (1963), "Deterministic Nonperiodic Flow", *J. Atmos. Sci.* **20**, 130.
Mandelbrot, B. (1982): *The Fractal Geometry of Nature*. Freeman, San Francisco.
Ott, E. (1981), "Strange Attractors and Chaotic Motions of Dynamical Systems", *Rev. Mod. Phys.* **53**. 655.
Ruelle, D. (1980), "Strange Attractors", *Math. Intelligencer* **2**, 126.

The Poincaré-Bendixson theorem is proved in:

Hirsch, M. W., and Smale, S. (1965): *Differential Equations, Dynamic Systems and Linear Algebra*. Academic Press, New York.

The area-preserving and the dissipative baker's transformation are discussed together with the Kaplan-Yorke conjecture in:

Kaplan, J. and Yorke, J. (1979) in H. O. Peitgen and H. O. Walther (eds.): *Functional Differential Equations and Approximation of Fixed Points.* Springer, Heidelberg-New York.

Lebowitz, J. L., and Pentrose, O. (1973), "Modern Ergodic Theory", *Phys. Today* **23** (2).

Russel, D. A., Hansen, J. D., and Ott, E. (1980), "Dimension of Strange Attractors", *Phys. Rev. Lett.* **45**, 1175.

Shaw, R. S. (1981), "Strange Attractors, Chaotic Behaviour and Information Flow", *Z. Naturforsch.* **36a**, 80.

The Kolmogorov entropy and the computation of Liapunov exponents are treated in:

Arnold, V. I., and Avez, A. (1974): *Ergodic Problems of Classical Mechanics.* Benjamin, New York.

Farmer, J. D. (1982a), "Information Dimension and the Probabilistic Structure of Chaos", *Z. Naturforsch.* **37a**, 1304.

Kolmogorov, A. N. (1959), *Dokl. Akad. Nauk. USSR* **98**, 527.

Pesin, Ya. B, (1977), *Usp. Mat. Nauk.* **32** (4) 55, engl. transl.: *Math. Surveys* **32** (4) 55.

Shannon, C. E., and Weaver, W. (1949): *The Mathematical Theory of Information* University of Ill. Press, Urbana.

Shimada, I., and Nagashima, T. (1979), "A Numerical Approach to Ergodic Problems of Dissipative Dynamical Systems", *Progr. Theor. Physics* **61**, 1605.

Wolf, A., and Vestano, J. A. (1986), "Intermediate Length Scale Effects in Liapunow Exponent Estimation", in G. Mayer-Kress (ed.): *Springer Series in Synergetics* **39**. Springer, Berlin-Heidelberg-New York-Tokyo.

References about the generalized dimensions and entropies of strange attractors are:

Ben-Mizrachi, A., Procaccia, I., and Grassberger, P. (1984), "The Characterization of Experimental (noisy) Strange Attractors", *Phys. Rev.* **29 A.** 975.

Eckmann, J. P., and Procaccia, I. (1986), "Fluctuation of Dynamical Scaling Indices in Non-Linear Systems", *Phys. Rev.* **34 A**, 659.

Eckmann, J. P., and Ruelle, D. (1985), "Ergodic Theory of Chaos and Strange Attractors", *Rev. Mod. Phys.* **57**, 617.

Farmer, J. D. (1982b), "Dimension, Fractal Measure and Chaotic Dynamics", in H. Haken (ed.): *Evolution of Order and Chaos.* Springer, Heidelberg-New York.

Fraser, A. M., and Swinney, H. L. (1986), "Independent Coordinates for Strange Attractors from Mutual Information", *Phys. Rev.* **33 A**, 1134.

Grassberger, P., and Procaccia, I. (1983a), "On the Characterization of Strange Attractors", *Phys. Rev. Lett.* **50**, 346.

Grassberger, P., and Procaccia, I. (1983b). "Estimation of the Kolmogorov Entropy from a Chaotic Signal", *Phys. Rev.* **29 A**, 2591.

Grassberger, P., and Procaccia, I. (1983 c), "Measuring the Strangeness of Strange Attractors", *Physica* **9 D**, 189.

Halsey, T. C., Jensen, M. H., Kadanoff, L. P., Procaccia, I., and Schraiman, B. I.
(1986), "Fractal Measures and their Singularities: the Characterization of Strange
Sets", *Phys. Rev.* **33 A**, 1141.

Liebert, W., Kaspar, F., and Schuster, H. G. (1987): "Proper Choice of the Time Delay
for the Analysis of Chaotic Time Series", to be published.

Mayer-Kress, G. (ed.) (1986): *Dimensions and Entropies in Chaotic Systems (Springer
Series in Synergetics* **32**). Springer, Berlin-Heidelberg-New York-Tokyo.

Pawelzik, K., and Schuster, H. G. (1987), "Generalized Dimensions and Entropies
from a Measured Time Series", *Phys. Rev.* **35 A**, 481.

Rössler, O. E. (1976), "An Equation for Continuous Chaos", *Phys. Lett.* **57 A**, 397.

Takahashi, Y., and Oono, Y. (1984), "Towards the Statistical Mechanics of Chaos,
Progr. Theoret. Phys. **71, 851.**

The reconstruction of an attractor from a time sequence is treated in:

Farmer, J. D. (1982 c), "Chaotic Attractors of an Infinite-Dimensional System",
Physica **4 D**, 366.

Mackey, M. C., and Glass, L. (1977), *Science* **197**, 287.

Packard, N., Crutchfield, J. P., Farmer, J. D., and Shaw, R. S. (1980), "Geometry from
a Time Series", *Phys. Rev. Lett.* **45**, 712.

Takens, F. (1981): *Lecture Notes in Math.* **898,** Springer, Heidelberg-New York.

Pictures of strange attractors and fractal boundaries:

Brolin, H. (1965), "Invariant Sets Under Iteration of Rational Functions", *Ark. for
Math.* **6**, 103.

Cvitanovic, P., and Myrheim, J. (1983), "Universality for Period-n Tuplings in Complex Mappings", *Phys. Lett.* **94 A**, 329.

Devaney, R. L. (1986): *An Introduction to Chaotic Dynamical Systems.* Benjamin-
Cummings, Menlo Park.

Douady. A., and Hubbard, J. H. (1982), "Iterations des Polynomes Quadratiques
Complexes", *Compt. Rend.* **294**, 93.

Fatou, P. (1919), *Bull. Soc. Math. France* **47**, 161; **48**, 33, 208 (1920).

Glass, L., Guevara, M. R., and Shrier, A. (1983), "Bifurcation and Chaos in a
Periodically Stimulated Cardiac Oscillator", *Physica* **7 D**, 89, Copyright 1981 by
A.A.A.S. Science, Vol. 214, pp. 1350–1353, 18. Dec. 1981.

Gumowski, I., and Mira, C. (1980), "Recurrences and Discrete Dynamic Systems",
Lect. Notes in Mathematics **809**, Springer, Berlin-Heidelberg-New York.

Hofstadter, D. R. (1979): *Gödel, Escher, Bach: an Eternal Golden Braid.* Basic Books,
New York.

Julia, G. (1918), "Mémoire Sur L'Iteration des Fonctions Rationelles", *J. Math. Pures et Appl.* **4**, 47.

Kawakami. H. (1984): *Strange Attractors in Duffing's Equation − 50 Phase Portraits of Chaos,* available from Univ. of Tokushima, Dept. of Electronics.

Klinker, T., Meyer-Ilse, W., and Lauterborn, W. (1984), "Period Doubling and Chaotic Behaviour in a Driven Toda Oscillator", *Phys. Lett.* **101 A**, 371.

Lauterborn, W., and Cramer, E. (1981), "Subharmonic Route to Chaos Observed in Acoustics", Phys. Rev. Lett. **47**, 1145.

Mandelbrot, B. B. (1980), "Fractal Aspects of the Iteration of $z \to \lambda z (1 - z)$ for Complex λ and z", *Ann. N. Y. Acad. Sci.* **357**, 249.

Martin, S., Leber, H., and Martienssen, W. (1984), "Oscillatory and Chaotic States of the Electrical Conduction in BSN Crystals", *Phys. Rev. Lett.* **53**, 303.

Peitgen, H. O., and Richter, P. H. (1984): *Harmonie in Chaos und Kosmos, and Morphologie komplexer Grenzen; Bilder aus der Theorie dynamischer Systeme.* Both catalogues can be obtained from: Forschungsschwerpunkt: Dynamische Systeme, Universität Bremen, D-2800 Bremen, F.R.G.

Peitgen, H. O., and Richter, P. H. (1986): *The Beauty of Fractals.* Springer, Berlin-Heidelberg-New York-Tokyo.

Pfister, G. (1984),"Period Doubling in Rotational Taylor-Couette Flow", *Proc. 2nd Int. Symp. on Appl. of Laser Anemometry to Fluid Mechanics.* Lisboa.

Prüfer, M. (1984), "Turbulence in Multistep Methods for Initial Value Problems", to be published in SIAM, *Journal of Appl. Math.*

Ruelle, D. (1980), "Strange Attractors", *Math. Intelligencer* **2**, 126.

Widom, M., Bensimon, D., Kadanoff, L. P., and Shenker, S. J. (1983), "Strange Objects in the Complex Plane", *J. Stat. Phys.* **32**, 443.

6   The Transition from Quasiperiodicity to Chaos

The onset of turbulence is discussed in:

Bergé, P. (1982), "Study of the Phase Space Diagrams Through Experimental Poincaré Sections in Prechaotic and Chaotic Regimes", *Physica Scripta* **T 1**, 71.

Brandstäter, A., Swift, J., Swinney, H. L., Wolf, A., Farmer D., Jen, E., and Crutchfield, P. (1983), "Low-Dimensional Chaos in a Hydrodynamic System", *Phys. Rev. Lett.* **51**, 1442.

Curry, J. H., and Yorke, J. A. (1978), "A Transition from Hopf Bifurcation to Chaos". *Lect. Notes in Math.* **668**, 48. Springer, New York.

Dubois, M., Bergé, P., and Croquette, V. (1982), "Study of the Steady Convective Regimes Using Poincaré Sections", *J. Phys. (Paris) Lett.* **43**, L 295.

Grebogi, C., Ott, E., and Yorke, J. A. (1983 a), "Are Three-Frequency Quasi-Periodic Orbits to be Expected in Typical Nonlinear Systems?" *Phys. Rev. Lett.* **51**, 339.

Hopf, E. (1942), "Abzweigungen einer periodischen Lösung von einer stationären Lösung eines Differentialgleichungssystems", *Math. Naturwiss. Klasse,* Sächs. Akademie der Wissenschaften, Leipzig, **94**, 1.

Landau, L. D. (1944), "On the Nature of Turbulence", *C. R. Dokl. Acad. Sci. USSR* **44**, 311.

Landau, L. D., and Lifshitz, E. M. (1959): *Fluid Mechanics.* Pergamon, Oxford

Libchaber, A., Fauve, S., and Laroche, C. (1983), "Two-Parameter Study of the Routes to Chaos", *Physica* **7D**, 73.

Marsden, J., and McCracken, M. (eds.) (1971): *the Hopf-Bifurcation Theorem and its Applications.* Springer, Heidelberg-New York.

Martin, S., Leber, H., and Martienssen, W. (1984), "Oscillatory and Chaotic States of the Electrical Conduction in BSN Crystals", *Phys. Rev. Lett.* **53**, 303.

Newhouse, S., Ruelle, D., and Takens, F. (1978), "Occurence of Strange Axiom-A Attractors near Quasiperiodic Flow on T^m, $m \leq 3$", *Commun. Math. Phys.* **64**, 35.

Ruelle, D., and Takens, F. (1971), "On the Nature of Turbulence", *Commun. Math. Phys.* **20**, 167.

Smale, S. (1967), "Differentiable Dynamical Systems", *Bull. Am. Math. Soc.* **13**, 747.

Swinney, H. L., and Gollub, J. P. (1978), "The Transition to Turbulence", *Phys. Today* **31 (8)**, 41.

Swinney, H. L., and Gollub, J. P. (1979), *J. Fluid Mech.* **94**, 103.

Swinney, H. L., and Gollub, J. P. (1980): *Hydrodynamic Instabilities and the Transition to Turbulence.* Springer, Heidelberg-New York.

The universal features in the transition from quasiperiodicity to chaos are investigated in the following articles:

Arnold, V. I. (1965), "Small denominators, I. Mappings of the Circumfence onto Itself", *Trans of the Am. Math. Soc.* **42**, 213.

Bak, P., Bohr, T., Jensen, M. H., and Christiansen, P. V. (1984), "Josephson Junctions and Circle Maps", *Solid State Commun.* **51**, 231.

Bak, P., and Jensen, M. H. (1985), "Mode Locking and the Transition to Chaos in Dissipative Systems", *Physica Scripta* **T 9**, 50.

Cvitanović, P. (1984): *Universality in Choas.* Adam Hilger, Bristol.

Cvitanović, P., and Söderberg, B. (1985 a), "Scaling Laws for Mode Lockings in Circle Maps", *Physica Scripta* **32**, 263.

Cvitanović, P., Jensen, M. H., Kadanoff, L. P., and Procaccia, I. (1985 b), "Renormalization, Unstable Manifolds and the Fractal Structure of Mode Locking", *Phys. Rev. Lett.* **55**, 343.

Guevara, M. R., Glass, L., and Shrier, A. (1981), "Phase Locking, Period Doubling Bifurcations and Irregular Dynmamics in Periodically stimulated Cardiac Cells", *Science* **214**, 1350.

Feigenbaum, M. J., and Haslacher, B. (1982), "Irrational Decimation and Path Integrals for External Noise", *Phys. Rev. Lett.* **49**, 605.

Feigenbaum, M. J., Kadanoff, L. P., and Shenker, S. J. (1982), "Quasiperiodicity in Dissipative Systems: A Renormalizations Group Analysis", *Physica* **5D**, 370.

Fein, A. P., Heutmaker, M. S., and Gollub, J. P. (1985), "Scaling at the Transition from Quasiperiodicity to Chaos in a Hydrodynamic System", *Phys. Scr.* **T9**, 79.

Greene, M. J. (1979), "A Method for Determining a Stochastic Transition", *J. Math. Phys.* **20**, 1183.

Hardy, G. H., and Wright, E. M. (1938): *Theory of Numbers.* Oxford Univ. Press, Oxford.

Hentschel, H. G. E., and Procaccia, I. (1983), "The Infinite Number of Generalized Dimensions of Fractals and Strange Attractors", *Physica* **8 D**, 435.

Hu, B. (1983). "A Simple Derivation of the Stochastic Eigenvalue Equation in the Transition from Quasiperiodicity to Chaos", *Phys. Lett.* **98 A**, 79.

Jensen, M. H., Bohr, T., Christiansen, P., and Bak, P. (1983 a), "Josephson Junctions and Circle Maps", *Solid State Commun.* **51**, 231.

Jensen, M. H., Bak, P., and Bohr, T. (1983 b), "Complete Devil's Staircase, Fractal Dimension and Universality of Mode-Locking Structures", *Phys. Rev. Lett.* **50**, 1637.

Jensen, M. H., Bak, P., and Bohr, T. (1984), "Transition to Chaos by Interaction of Resonances in Dissipative Systems I, II", *Phys. Rev.* **A 30**, 1960, 1970.

Jensen, M. H., Kadanoff, L. P., Libchaber, A., Procaccia, I., A Libchaber, and Stavans, J. (1985), "Global Universality at the Onset of Chaos: Results of a Forced Rayleigh − Bénard Experiment", *Phys. Rev. Lett.* **55**, 2798.

Martin, S., and Martienssen, W. (1986), "Circle Maps and Mode Locking in the Driven Electrical Conductivity of Barium Sodium Niobate Crystals", *Phys. Rev. Lett.* **56**, 1522.

Rand, D., Ostlund, S., Sethna, J., and Siggia, E. D. (1982), "Universal Transition from Quasiperiodicity to Chaos in Dissipative Systems", *Phys. Rev. Lett.* **49**, 132, and *Physica* **8 D**, 303 (1983).

Shenker, S. J. (1982), "Scaling Behaviour in a Map of a Circle into Itself: Empirical Results", *Physica* **5 D**, 405.

Vul, E. B., and Khanin, K. M. (1982), "On the Unstable Manifold of the Feigenbaum Fixed Point", *Usp. Math. Nauk.* **37**, 173.

Yeh, W. H., He, D. R., and Kao, Y. H. (1984), "Fractal Dimension and Self-Similarity of the Devil's Staircase in a Josephson Junction Simulator", *Phys. Rev. Lett.* **52**, 480.

Additional transition scenarios are discussed in:

Campbell, D., and Rose, H. (eds.) (1983), "Order in Chaos: Proceeding of the International Conference in Los Alamos", *Physica* **7 D.**

Gollub, J. P., and Benson, S. V. (1979): "Phase Locking in the Oscillations Leading to Turbulence", in H. Haken (ed.): *Pattern Formation and Pattern Recognition.* Springer, Heidelberg-New York.

Grebogi, C., Ott, E., and Yorke, J. A. (1983 b), "Crises, Sudden Changes in Chaotic Attractors and Transients to Chaos", *Physica* **7 D**, 181.

Swinney, H. L. (1983), in D. Campbell and H. Rose (eds.): "Order in Chaos. Proceedings of the International Conference in Los Alamos". *Physica* **7 D**, 1.

7 Regular and Irregular Motion in Conservative Systems

Besides the classic books by Poincaré (1982), Arnodl and Avez (1968), Moser (1973), and Balescu (1975), which we mentioned (together with the article by Lorenz (1963) on dissipative systems) in the list of reference of the introduction, there are several other excellent reviews and monographs on the irregular behavior of Hamiltonian systems:

Arnold, V. I. (1978): *Mathematical Methods of Classical Mechanics.* Springer, Heidelberg-New York.

Berry, M. V. (1978), in S. Jorna (ed.), "Topics in Nonlinear Dynamics", *Am. Inst. Phys. Conf. Proc.,* Vol. **46.**

Helleman, R. H. G. (1980), "Self-Generated Chaotic Behaviour in Nonlinear Mechanics", in E. G. D. Cohen (ed.): *Fundamental Problems in Statistical Mechanics, Vol.* **5.** North-Holl. Publ., Amsterdam.

Lichtenber, A. J., and Liebermann, M. A. (1982): Regular and Stochastic Motion. Springer, Heidelberg-New York.

The KAM theorem is proved in:

Arnold, V. I. (1963), "Small Denominators II, Proof of a theorem of A. N. Kolmogorov on the Preservation of Conditionally-Periodic Motions Under a Small Perturbation of the Hamiltonian", *Russ. Math. Surveys* **18,** 5.

Kolmogorov, A. N. (1954), "On Conservation of Conditionally-Periodic Motions for a Small Change in Hamilton's Function", *Dokl. Akad. Nauk. USSR* **98,** 525.

Moser, J. (1967), "Convergent Series Expansions of Quasi-Periodic-Motions", *Math. Ann.* **169,** 163.

More references on this topic can be found in the book by Lichtenberg and Liebermann (1982).

Moser's twist map is discussed in:

Moser, J. (1973): *Stable and Random Motions in Dynamical Systems.* Princeton University Press.

The Poincaré-Birkhoff theorem can be found in:

Birkhoff, G. D. (1935), "Nouvelles Recherches Sur les Systems Dynamiques", *Mem. Pont. Acad. Sci. Novi Lyncaei* **1,** 85.

Some applications and examples are given in:

Anosov, D. V. (1969), "Geodesic Flows on Closed Riemannian Manifolds with Negative Curvature", *Am. Math. Soc.*

Arnold, V. I. (1964), "Instability of Dynamical Systems with Several Degrees of Freedom", *Sov. Math. Dokl.* **5**, 581. In this article the so-called Arnold diffusion is introduced.

Aubry, S., and André, C. (1980), in: *Proceedings of the Israel Physical Society, Vol.* **3**, p. 133.

Benettin, G., and Streleyn, J. M. (1978), "Numerical Experiments on the Free Motion of a Point Mass, Moving in a Plane Convex Region: Stochastic Transition and Entropy", *Phys. Rev.* **17 A**, 773.

Bunimovich, L. A. (1979), "On the Ergodic Properties of Nowhere Dispersing Billiards", *Commun. Math. Phys.* **65**, 295.

Gustavson, F. (1966), "On Constructing Formal Integrals of a Hamiltonian System near an Equilibrium Point", *Astron. J.* **71**, 670.

Hénon, M., and Heiles, C. (1964), "the Applicability of the Third Integral of the Motion: Some Numerical Experiments", *Astron. J.* **69**, 73.

Sinai, Y. G. (1970), "Dynamical Systems wiht Elastic Reflections", *Russ. Math. Surv.* **25**, 137.

The universal behavior of period doubling for two-dimensional conservative maps is treated in the review article by Helleman which is mentioned under the general sources of the introduction and in:

Bountis, T. C. (1981), "Period-Doubling Bifurcations and Universality in Conservative Systems", *Physica* **3 D**, 577.

Greene, J. M., McKay, R. S., Vivaldi, F., and Feigenbaum, M. J. (1981), "Universal Behaviour of Area-Preserving Maps", *Physica* **3 D**, 468.

There is a rapidly growing number of articles on renormalization schemes for area-preserving maps, some examples are:

Escande, D. F., and Doveil, F. (1981), "Renormalization Method of Computing the Threshold of the Large Scale Stochastic Instability in Two Degrees of Freedom Hamiltonian Systems", *J. Stat. Phys.* **26**, 257.

Kadanoff, L. P. (1981), "Scaling for a Critical K.A.M. Trajectory", *Phys. Rev. Lett.* **47**, 1641.

McKay, R. S. (1983), "A Renormalization Group Approach to Invariant Circles in Area-Preserving Maps", *Physica* **7 D**, 283.

McKay, R. S., Meiss, J. D., and Percival, I. C. (1984), "Stochasticity and Transport in Hamiltonian Systems", *Phys. Rev. Lett.* **52**, 697.

Shenker, S. J. and Kadanoff, L. P. (1982), "Critical Behaviour of a K.A.M. Trajectory: I. Empirical Results", *J. Stat. Phys.* **27**, 631.

More on the classification of chaos in conservative systems can be found in:

Lebowitz, J. L. (1972) in S. A. Rice, K. F. Freed, J. C. Light (eds.): *Proc. of the Sixth IUPAP Conf. on Statistical Mechanics.* Univ. of Chicago Press, Chicago.
Lebowitz, J. L., and Penrosé, O. (1973), "Modern Ergodic Theory", *Phys. Today* **26** (2).

8   Chaos in Quantum Systems?

The problem of quantization of classical chaotic systems was already recognized by:

Einstein, A. (1917): "Zum Quantensatz von Sommerfeld und Epstein", *Verh. Dtsch. Phys. Ges.* **19**, 82.

There are several excellent reviews on the problem of stochasticity in quantum system:

Berry, M. V. (1983), in R. H. G. Helleman, and G. Ioos (eds.): *Chaotic Behaviour of Deterministic Systems. Les Houches Summer School 1981.* North-Holland, Amsterdam.
Casati, G. (1982), in H. Haken (ed.): *Evolution of Order and Chaos.* Springer, Heidelberg-New York.
Chirikov, B. V. (1979): "A Universal Instability of Many-Dimensional Oscillator Systems", *Phys. Rep.* **52**, 463.
Zaslavsky, G. M. (1981): "Stochasticity in Quantum Systems", *Phys. Rep.* **80**, 157.

More on laser photochemistry can be found in:

Ben Shaul, A., Haas, Y., Kompa, K. L., and Levine, R. D. (1981): *Laser and Chemical Change.* Springer, Heidelberg-New York.

The quantized version of Arnold's cat map was first studied by:

Hannay, J. H., and Berry, M. V. (1980), "Quantization of Linear Maps on a Torus, Fresnel Diffraction by a Periodic Grating", *Physical* **1D**, 267.

The quantum particle in a stadium was investigated by:

McDonald, S. W., and Kaufman, A. N. (1979), "Spectrum and Eigenfunction for a Hamiltonian with Stochastic Trajectories", *Phys. Rev. Lett.* **42**, 1189.

The kicked quantum rotator is discussed in the following articles:

Casati, G., Chirikov, B. V., Izraelev, F. M., and Ford, J. (1977), in G. Casati, and J. Ford (eds.): *Stochastic Behaviour in Classical and Quantum Hamiltonian Systems. Lecture Notes in Physics* **93**, Springer, Heidelberg-New York.

Grempel, D. R., Fishman, S., and Prange, R. E., (1982), "Localization in an Incommensurate Potential: An Exactly Solvable Model", *Phys. Rev. Lett.* **49**, 833.

Fishmann, S., Grempel, D. R., and Prange, R. E. (1982), "Chaos Quantum Recurrences and Anderson Localization", *Phys. Rev. Lett.* **49**, 509.

Hogg, T., and Huberman, B. A. (1982), "Recurrence Phenomena in Quantum Dynamics", *Phys. Rev. Lett.* **48**, 711.

Hogg, T., and Huberman, B. A. (1983), "Quantum Dynamics and Nonintegrability", *Phys. Rev.* **28 A**, 22.

Izraelev, F. M., and Shepelyanskii, D. L. (1980), "Quantum Resonance for a Rotator in a Nonlinear Periodic Field", *Theor. Math. Phys.* **43**, 417.

Peres, A., (1982), "Recurrence Phenomena in Quantum Dynamics", *Phys. Rev. Lett.* **49**, 1118.

Schuster, H. G. (1983), "Absence of Quasiperiodicity in a Kicked Quantum Rotator and a New Delocalization Transition", *Phys. Rev.* **28 B**, 381.

The electron localization problem has been investigated for the general case by:

Anderson, P. W. (1958), "Absence of Diffusion in Certain Random Systems", *Phys. Rev.* **103**, 1492.

and especially for the one-dimensional case by:

Ishii, K. (1973), "Localization of Eigenstates and Transport Phenomena in One-Dimensional Disordered Systems", *Progr. Theor. Phys. Suppl.* **53**, 77.

Some more recent references on the question of chaos in quatum systems are:

Berry, M. V., and Tabor, M. (1977), "Level Clustering in the Regular Spectrum", *Proc. Roy. Soc. (London)* **A 356**, 375.

Blümel, R., and Smilansky, U. (1984), "Quantum Mechnical Suppression of Classical Stochasticity in the Dynamics of Periodically Disturbed Surface Electrons", *Phys. Rev. Lett.* **52**, 137.

Bohigas, O., Giannoni, M. J., and Schmit, C. (1984), "Characterization of Chaotic Quantum Spectra and Universality of Level Fluctuation Laws", *Phys. Rev. Lett.* **52**, 1.

Casati, G., (ed.) (1985): *Chaotic Behavior in Quantum Systems. Proc. 1983 Como Conf. on Quatum Chaos.* Plenum Press, New York.

Gutzwiller, M. C. (1983), "Stochastic Behaviour in Quantum Scattering", *Physica* **7 D**, 341.

Seligman, T. H., and Nishioka, H. (eds.) (1986): *Quantum Chaos and Statistical Nuclear Physics.* Springer, Berlin-Heidelberg-New York.

Outlook

General:

Abraham, N. B., Gollub, J. P., and Swinney, H. L. (1984), "Testing Nonlinear Dynamics", *Physica* **11 D**, 252.

Benedek, G., Bilz, G., and Zeyher, R. (1984), "Statics and Dynamics of Nonlinear Systems", *Springer Series in Solid-State Sciences* **47**, Springer, Berlin-Heidelberg-New York.

Berry, M. V. (1978), in S. Yorna (ed.), "Topics in Nonlinear Dynamics", *Am. Inst. Phys. Conf. Proc.,* Vol. **46**.

Kaneko, K. (1986): *Collapse of Tori and Genesis of Chaos in Dissipative Systems.* World Scientific Publishing Co., Singapore.

Coupled Systems:

Bishop, A. R., Fesser, K., Lomdahl, P. S., Kerr, W. C., Trullinger, S. E., and Williams, M. B. (1983), "Coherent Spatial Structures Versus Time Chaos in a Perturbed Sine-Gordon System", *Phys. Rev. Lett.* **50**, 1095.

Kuramoto, Y.(1977), "Chemical Waves and Chemical Turbulence", in H. Haken (ed.): *Synergetics.* Springer, Berlin-Heidelberg-New York.

Vidal, C., and Pacault, A. (1981): *Nonlinear Phenomena in Chemical Dynamics.* Springer, Berlin-Heidelberg-New York.

Winfree, A. T. (1980), "The Geometry of Biological Time", *Biomathematics* **8**, Springer, Berlin-Heidelberg-New York.

Turbulence:

Ruelle, D. (1983), "Five Turbulent Problems", *Physica* **7 D**, 40.

Quantum Systems:

Ackerhalt, J. R., Milonni, P. W., and Shih, M. L. (1985), "Chaos in Quantum Optics", *Phys. Rep.* **128**, 207.

Graham, R. (1984), "Chaos in Lasers", in E. Frehland (ed.): *Synergetics — From Microscopic to Macroscopic Order.* Springer, Berlin-Heidelberg-New York.

Elgin, J. N., and Sarkar, S. (1984), „Quantum Fluctuations and the Lorenz Strange Attractor", *Phys. Rev. Lett.* **52**, 1215.

Misra, B., and Prigogine, I. (1980): "One the Foundations of Kinetic Theory", *Suppl. Progr. Theor. Physics* **69**, 101.

Random Numbers:

De Long, H. (1970): *Randomness and Gödel's Incompleteness Theorem.* Addison-Wesley, Reading, MA.

Biology:

Hess, B., and Markus, M. (1984), "Time Pattern Transitions in Biochemical Processes", in E. Frehland (ed.): *Synergetics — From Microscopic to Macroscopic Order.* Springer, Berlin-Heidelberg-New York.

Schuster, P. (ed.) (1984), "Stochastic Phenomena and Chaotic Behaviour in Complex Systems", *Springer Series in Synergetics* **21**, Springer, Berlin-Heidelberg-New York.

Cellular Automata:

Wolfram, S. (1985), "Undecidability and Intractability in Theoretical Physics", *Phys. Rev. Lett.* **54**, 735.

Frisch, U., Hasslacher, B., and Pomeau, Y, (1986), "Lattice-Gas Automata for the Navier Stokes Equation", *Phys. Rev. Lett.* **56**, 1505.

Farmer, D., Toffoli, T., and Wolfram, S. (1984): "Cellular Automata", *Physica* **10 D**.

Subject Index